Scribe Publications
MAKING BABIES

Theresa Miller has twenty years' experience in the media. Her first job was as a reporter for Channel Nine News in Adelaide. She then moved to Europe for six years, where she worked as a producer for *Good Morning Britain*, BskyB's 24-hour news channel, and The European Business Channel, and reported for the CNN World Report. Since her return to Sydney, Theresa has freelanced as a reporter, presenter, and producer for Channels Nine, Seven, SBS, and ABC TV.

In 2000, Theresa was a media adviser to the Sydney lord mayor during the Olympics. Since then she's worked as a media trainer and journalism lecturer, and produces for ABC Radio National's 'Life Matters'.

Theresa has also appeared in numerous theatre productions, TV soaps, and commercials. She and her husband Stuart live in Sydney with their IVF daughter, Zoë, and their recently born 'home grown' baby, Sienna.

To my mother, Alina

Personal **IVF** *Stories*

making babies
THERESA MILLER

SCRIBE
Melbourne

Scribe Publications Pty Ltd
PO Box 523
Carlton North, Victoria, Australia 3054
Email: info@scribepub.com.au

First published by Scribe 2007

Copyright © Theresa Miller 2007

All rights reserved. Without limiting the rights under copyright reserved above, no part of this publication may be reproduced, stored in or introduced into a retrieval system, or transmitted, in any form or by any means (electronic, mechanical, photocopying, recording or otherwise) without the prior written permission of the publisher of this book.

Designed and typeset in 12/15.75 pt Granjon by the publisher
Cover design by Nada Backovic Designs
Printed and bound in Australia by Griffin Press

National Library of Australia
Cataloguing-in-Publication data

Miller, Theresa.
 Making babies : personal IVF stories.

ISBN 9781921215469 (pbk.).

 1. Fertilization in vitro, Human - Australia.
 2. Infertility - Treatment - Australia. I. Title.

618.1780599

www.scribepublications.com.au

Contents

Acknowledgements vii
Introduction 1
CHAPTER 1   Touched by an Angel   7
CHAPTER 2   Ugly Brown Eggs   31
CHAPTER 3   Sometimes It Takes Three to Make a Baby   48
CHAPTER 4   Too Little, Too Late   64
CHAPTER 5   Towards the Light   75
CHAPTER 6   Desperately Seeking Daddy   87
CHAPTER 7   Defying Mother Nature   100
CHAPTER 8   Sister Versus Sister   113
CHAPTER 9   Double Life   123
CHAPTER 10   From Serbia with Love   135
CHAPTER 11   The Gift of Giving   143
CHAPTER 12   Test-Tube Grown-up   154
CHAPTER 13   Father Pride   160
CHAPTER 14   Single-Minded   173
Glossary 187

Acknowledgements

This book would not have been possible without the people who opened their hearts and minds and shared their important stories, including those whose interviews did not make it to publication. I am also very grateful to the on-line community of essentialbaby.com.au for responding so enthusiastically to my ad for IVF stories. Their overwhelming response confirmed to me that this book was indeed needed.

Thank you to my friend and coach, Bernadette Schwerdt, whose encouragement and inspiration helped me get this idea out of my head and onto the page.

To my husband Stuart Ziegler, I am forever grateful for the support, space, and love he has given me to write. His diverse skills as 'daddy-day-care' and rigorous book editor are unrivalled and much appreciated.

Thank you also to my agent, Benython Oldfield, for taking a punt on me, and to Henry and Margot Rosenbloom at Scribe for jumping so quickly and keenly on this project.

Thanks also to my writing cheerleader, Libby-Jane Charleston, and to Katherine Roche at Sydney IVF for supporting this concept and coming up with the book's title. I'm also very grateful to Stephanie Miller for promptly and efficiently transcribing my recorded interviews.

And, last but not least, thank you to my miracle babies: Zoë, whose IVF conception inspired this anthology; and my second, surprise 'home grown' pregnancy, which kept me company by wriggling inside me during the latter stages of editing. It was a photo-finish between her due date and my publishing deadline.

Introduction

I was conceived accidentally by a couple of teenagers in the back seat of a 1960s pink Ford Zephyr at the drive-in movies in Adelaide. I don't know what was playing that night, but it obviously didn't capture my parents' attention. Four months later, my Catholic grandparents marched their disgraced children down the aisle.

Except for my grinning father wearing a tight, borrowed suit, my eighteen-year-old mother and the rest of the family looked grim-faced in the wedding photos. My parents went on to have my little sister three years later. Their marriage lasted fifteen years, which is not a bad track record for a shotgun wedding.

And so it was that my mother warned me not to make the same mistakes she'd made. 'Don't get married young, see the world, go to university, have a career, have lots of boyfriends before you settle down and, most importantly, don't get pregnant accidentally!'

Dutifully, I followed my mother's instructions. I went to university and studied journalism, landed a job as a TV reporter, worked in London and Europe for six years, lived with my violinist boyfriend in Switzerland, and travelled the world.

When I met my husband-to-be, Stuart Ziegler, in Sydney, I was 31 and ready to settle down. Within a few months, I fell pregnant accidentally. I was excited, but Stuart wasn't so thrilled. Our relationship was still new and he was worried about how he'd support us. My mother's words were ringing in my head: 'Don't ever make a man marry you because you're pregnant.' So, with a heavy heart, I had a termination. This was a decision that we both came to deeply regret.

Six months later, Stuart and I were married. I threw my contraceptive pill away and we tried in earnest to start a family. Nothing happened after the first year, but I wasn't too worried. I was working as a TV reporter and travelling often. It was probably just bad timing, I told myself. After the second year, I began to worry I'd damaged my fallopian tubes, somehow, with the termination. But tests revealed that everything was fine.

By the third year, the strain was taking its toll on our marriage, and I blamed Stuart for 'making me have an abortion'. We began to argue more than we were having sex. By the fourth year, family and friends stopped asking about the 'pitter-patter of little feet'. When I heard about friends falling pregnant easily I'd smile and congratulate them, and go home and cry.

I started to investigate IVF, but the only books I found were technical manuals and a devastating memoir by a woman who tried unsuccessfully for years and suffered terrible side-effects from the drugs.

At first, I stubbornly rejected IVF, saying, 'We've conceived once naturally; we can do it again!' Instead we spent a fortune on acupuncture, naturopaths, Chinese herbalists, spiritual healers, and ayurvedic medicine. By now my sense of humour was drying up and, according to my doctor, so were my eggs.

Around the time of my 37th birthday I met a woman at a party who told me she'd just had twins using IVF. When I told her my age, and that we'd been trying to conceive for five years, she said, 'For God's sake, woman, get yourself down to the Baby Factory and get on the IVF program. You've got no time to lose!'

So that's exactly what we did. After talking to the nurses at

Sydney IVF, I threw down my Visa card and said, 'Book us in.' At last I felt like we were doing something proactive. Every morning Stuart would inject me in the bottom and, except for one jab, which made me feel like my legs were crawling with ants, I didn't have any adverse reactions to the drugs.

I didn't tell anyone at work what we were doing, but every morning I felt buoyed by my secret when I logged on to my computer with the password 'Zoë Ziegler'.

Harvest or egg pick-up day was the first anniversary of September 11. As I placed my legs in stirrups and winced while the doctor extracted eggs with a long needle from my pumped-up ovaries, I wondered what sort of world I would be bringing a child into. But the human instinct to procreate seems to override logic, good sense, and even fear.

My pride at producing the grand total of nineteen eggs — as if I was a prize-winning chook — was dashed the next day when only three fertilised. I couldn't help wondering whether my crusty old eggs were to blame or my husband's lazy sperm.

At Sydney IVF they grow the fertilised egg for five days until it's a multi-celled blastocyst, before transferring it into the mother's womb. It seemed surreal that while we were at work or out to dinner, our 'offspring' were growing in a petri dish in the city.

Every day Stuart would ring the lab to see how 'the little guys' were doing. In the meantime, I tried to convince Stuart we should have two embryos, rather than one, transferred to increase my chances, even though our doctor had warned us we could end up with twins. I left a letter on his desk headlined: 'Ten Reasons Why We Should Have Twins' followed by bullet points. Stuart still laughs about it today, and wishes he'd kept that paper to remind me whenever I complain what a handful one child is.

As it turned out, we didn't have the twin option. According to the lab, one blastocyst was way out in front as an 'A' grade specimen, which meant the cells were dividing rapidly while the other two were growing more slowly. They recommended transferring the good one and freezing the other two as back-up. As it turned out, the slower blastocysts stopped dividing and simply disintegrated before they even got to the freezer. I was devastated. The doctor tried to reassure me. 'It's not every day I get to transfer such a good-looking blastocyst,' he said.

I'll never forget looking down the microscope at what we nicknamed the 'blasting blastocyst' that was to became Zoë Ziegler. After the doctor had transferred the fertilised egg into my uterus, I asked him if I should go home and put my legs up, so it wouldn't fall out. He laughed. 'There are women out there who have no idea they have a five-day-old embryo growing inside them, and they're drinking champagne and dancing all night. Now it's simply up to that embryo whether if wants to become a baby or not.'

Somehow, I found that strangely reassuring. For all its incredible technology, IVF still has to leave room for the magic and mystery of creation.

Zoë is Greek for 'life'. Today, as I look at my beautiful, bright, and bubbly three-year-old daughter, I don't just marvel at the wonder of IVF; I marvel at the wonder of her and all children.

Why does new life sometimes spring unbidden from a once-off romp in the back of a car and at other times refuse to blossom despite years of yearning? My newfound awe sent me on a quest to interview other people who'd also experienced IVF. I sought both men's and women's personal stories. As it turned out, it was mostly women who responded. I was touched by their open-hearted and candid stories. Together we sat in their kitchens or on their sofas, and laughed and wept at their journeys.

Not all the stories in this book have happy endings like mine. Some have given up IVF after years of trying without success; others are still on the treadmill.

After countless miscarriages, one woman finally gave birth to a baby which tragically died weeks later from a rare congenital disease. Another couple gave birth to twins after a friend donated her eggs, while a mother of two impulsively donated her eggs to a stranger.

Women also tell of enduring personal tragedies in their quest for a child; while one woman mourned her brother's suicide, another was dumped by her partner in the middle of her IVF cycle. Neither gave up their dream of becoming a mother.

I also spoke to a remarkable young woman who was the product of one of the earliest IVF programs. At school she was teased and called a 'test-tube baby'; now she's an ambassador for an infertility network.

Assisted reproductive technology has also made it possible for gays and singles to be parents, too. In this book, a gay male couple and a single woman in her forties share their stories of baby hunger.

All these memoirs are very different. All display courage, determination, vulnerability, love, and proof that the desire for a baby is bigger than us all.

Postscript
As I wrote this book, my pregnant belly pressed against the desk. After Zoë turned three, we decided not to do IVF again and to be content with one child. I gave away the high chair, the pram, and my maternity clothes. A month later my hands shook as I held the pregnancy test and looked at the two red lines showing a positive result. Our second daughter, Sienna, the home-grown type, was born in early 2007.

CHAPTER 1

Touched by an Angel
–Kirrily's story

It took six miscarriages plus the death of their premature baby to convince Kirrily and Steve to try IVF and genetic testing. Although Kirrily fell pregnant easily, a rare chromosomal problem on her husband's part meant that every two out of three embryos they conceived were abnormal. It was their intense but brief experience as parents of baby Ella that gave them the determination to keep trying. Their painful journey changed both of them immeasurably, and put Kirrily in touch with her intuition.

My husband, Steve, and I were childhood sweethearts. We got together in 1993 when I was seventeen and he was twenty. We were married six years later, and never really talked about having kids until we were absolutely ready. Steve is one of those people who prefers to plan rather than dream. A year after we married, I fell pregnant naturally on our first attempt. I must admit I felt pretty cocky, and Steve was positively strutting.

Unfortunately, that all went sour at eight or nine weeks. I had a dream I was holding my tummy and it expanded really quickly, like fast-rising bread in the oven. Then it just went 'BANG'. I woke up and thought something had gone wrong. I didn't feel pregnant anymore; it was instinctual rather than logical.

I went to the doctor and he said, 'It's just jitters.' I was twenty-four, and he treated me as if I were too young to know what I was talking about. 'You can't possibly know,' he said. He waved me off with his hand and turned his head away. He ignored my concerns but, sure enough, I started spotting about three weeks later when I was nearly thirteen weeks pregnant. Because I had no pain it was termed a 'missed' miscarriage. A doctor in emergency at the local hospital carried out my D&C. When I went back to my GP for a follow-up check he was on leave. Fortunately I saw a very compassionate female doctor. When my GP came back he didn't even bother to contact me. Needless to say, I never went back to see him. It was a really horrible experience and a pretty cruel first pregnancy.

In some respects, that first experience prepared me for what was to come. And I've slowly learned over the years to trust my instincts, rather than listen to people's 'don't be silly' comments. That dream about the rising and exploding tummy has become a recurring one, and I've had it several times before I've miscarried.

The first miscarriage was such a blow to Steve and me. We didn't really talk about it. Instead, we just let it fester and eat us away — so much so that we separated for two weeks. Fortunately, we went to marriage counselling, first separately and then as a couple for six months. The counsellor helped us realise that the strain on our marriage stemmed from our grief over the miscarriage. We didn't want to believe it at first. We kept saying, 'Oh no, we're over that now.' It was hard to admit, 'Yes, actually, this deep anger and resentment between us is because of the miscarriage.' For my part I felt cheated, as if my body had let me down. I was really hard on myself. Counselling was very cathartic for Steve. In one session he broke down and said, 'I feel like I've failed.'

I didn't try to get pregnant again for a while. The miscarriage scared us off. At that stage we didn't know anything about chromosomal abnormalities. We just put it down to 'one of those things', because that's what everyone kept telling us. In hindsight, we were too passive and let other people's comments influence us, such as, 'It wasn't meant to happen … The baby must have had something wrong with it … You wouldn't have wanted to have it anyway.' That was a trap, because if you hear those things long enough you begin to buy into it. We became complacent instead of immediately investigating the cause of the miscarriage.

It wasn't until the beginning of 2002 that we tried to conceive again. Bingo, I got pregnant straightaway, but that also ended in a miscarriage at about five weeks. I then had another miscarriage a few months later. I thought, *This is crazy; there must be something wrong with me*, because there was no problem getting pregnant. I saw an obstetrician and he said, 'You are 26, you've got plenty of time.' I'd heard this talk before and thought, *I am not leaving here without some answers*. That's when I began to assert myself.

'I want an investigation,' I said.

'Well, you're still very young, and we don't normally do tests until you've had a few miscarriages and …'

'Well, how's three miscarriages in a row?' I said. 'Is that enough for you?'

Finally, the doctor sent Steve and me for blood tests. A letter came back saying, lo and behold, they had found something in Steve's results called chromosomal translocation. We were referred to a specialist at Monash Genetics who could tell us more about it.

Chromosomal translocation happens when two pieces of DNA are swapped around. It can happen to either gender. People like Steve, who have no physical or mental disabilities, have what's called a balanced translocation — which means the genetic information is complete, but the chromosomes are in the wrong order. Consequently, two out of three of the embryos we conceive will be abnormal and miscarry. Steve can also pass the defect on and have a baby who is outwardly normal, but with a couple of mixed-up chromosomes, who in turn can have the same problem when trying to conceive and could pass on something to their offspring.

It was also possible I could carry and give birth at term to a baby who was severely mentally and physically disabled, and would only live a short time. There is no way to really know during pregnancy unless you have the amniocentesis test at twelve weeks, but that also increases the risk of miscarriage.

Steve was heartbroken when we first got the news. Even now he struggles with it. He suffered a great deal of guilt. He tortured himself, knowing I'd been through the physical pain of three miscarriages and that we'd both endured emotional loss. He was so depressed at first, he could barely get off the couch. He didn't even want contact with our little niece and nephew, whom he loves and who adore him. Normally, they were the highlight of our weekend; but, after the news, he stopped visiting them. Steve felt the situation was hopeless because the problem was in every cell of his body. He believed that, even if he did homeopathy or stood on his head, it wouldn't change a thing; we were still lumped with those rotten odds. Consequently, he didn't do anything to improve his sperm quality or general health with vitamins or exercise or anything, until much later.

We had a few friends offer their sperm, but fortunately they only suggested it to me. One insensitive, dim-witted friend said, 'Why can't Steve's brother just give Kirrily his sperm?' Those sorts of comments really hurt Steve. It's horrifying what people say without realising the impact.

The geneticist at Monash suggested we try IVF plus a new procedure called PGD, or Pre-implantation Genetic Diagnosis, in which, three days after the egg has been fertilised in a petri dish, they inject it with a solution to separate the cells, making them easier to extract and test. In my view, it's way too much meddling, but it was the only way we could have a stress-free pregnancy and not worry about whether the baby had the condition or not.

However, the geneticist told us doing IVF put us in the same risk-category of miscarriage as we had now. So we thought, *Okay, we'll stick with those odds, and keep trying naturally.* We didn't really have the money for IVF and the PGD. So we just kept going. I got pregnant again in September 2002 and we lost it at six weeks.

It's been hard for me to hear people's words of comfort because I don't feel like they really understand. I also struggle with what to say to people when I've heard they've had a miscarriage, too. I never assume to know how they feel. I just say, 'I am really sorry and I am always here if you want to talk.' Each miscarriage is damaging and difficult in its own way. I started to think, *Is this my pattern now? Am I never going to have a baby?* Then the desperation sets in and you think, *I have just had another miscarriage; it must have been because I was stressing.* It's a vicious circle.

After the fourth miscarriage, I didn't get pregnant again for months, and I thought that was unusual. We were barely coping with the miscarriages, but then suddenly not falling pregnant at all was an added concern. I decided to see a homeopath because I'd put on about ten kilograms I couldn't lose. She diagnosed me with candida, which can sometimes prevent conception and can be caused by multiple pregnancies and antibiotics. Following her advice, I did the candida diet for three months and took homeopathic medicine until I was back on track.

Our fifth pregnancy produced Ella. I had a dream about her a few days before we conceived. I dreamed that a little girl who looked just like me was playing in a cot and she was completely unconcerned, almost blasé about my presence, while I was absolutely enraptured by her.

My immediate thought was, *Oh, no, this pregnancy is not going to work; it'll end just like the rest.* Except, in the back of my mind, I heard what sounded like someone clanging saucepan lids to let me know they were there.

I first became aware of Ella's presence in November 2003. I didn't know who it was, but it was like knowing there was someone down the other end of the house. It was like when your husband is home, you can't see him, but you can tell he's there. That's what it was like with her, except she was a lot closer, as if she was right next to me. She had a very strong presence and a big personality. She was pleasant and happy and brought me a lot of happiness.

From about five weeks into the pregnancy I knew it was a girl; there was no doubt in my mind. When we had the ultrasound much later on, the radiographer asked, 'Do you want to know what sex it is?'

'It is a girl,' I said.

'You're right; it's a girl.'

'Yeah, I know.'

As soon as I found out I was pregnant, I gave four weeks' notice at my secretarial job. I felt more confident about this pregnancy than any of the others, and wanted to give it my full attention.

But just a week later, I honestly thought we'd lost her, because I went to the toilet before bed one night, and rolling around in my undies was something that looked like a dice of dry blood. It was really bizarre. Steve and I lay on our bed and had a huge sob and I said, 'I can't do this anymore.'

We were so relieved when the doctor confirmed I was still pregnant. And then on the eighth week, I was at work when I started bleeding. By the time I got to the toilet I was haemorrhaging bright, red blood. It was the scariest thing. My world felt as if it was crashing down around me. I thought, *I have just quit a well-paid, good job to make a go of this pregnancy, and I've just lost it down the toilet.*

I rang Steve and said, 'Come home, we've lost it, come home.' I rang and told my doctor, and he said, 'It does sound like it, yes.'

That trip home was just horrible, and the colour of the blood scared me to death. By the time I got home I had filled a pad with blood. We didn't even cry. We were too numb.

That was on the Monday, and I had had an ultrasound scheduled for the Wednesday. The next day I stayed home, and just bled and bled and felt gross. I wasn't sure if I should bother going to the ultrasound the next day, but they asked me to come in anyway. The next morning I rang my sister.

'I don't think I'll be able to drive because I'm feeling pretty weak and sick.'

'Don't worry, I'll come with you,' she said. 'I'll organise the kids and come and get you.'

When I got to the clinic I had to go through yet another miscarriage story and the history of all my failed pregnancies.

As I lay on the table for the ultrasound I said, 'You know, the strange thing is, I don't think it has all come out yet because there haven't been any clots or pain, just blood.'

'You're right. It hasn't,' the radiographer said.

I was staring at the ceiling because I always hated looking at the ultrasound in case she wasn't there. I looked over at him when he said that, and he was grinning.

I looked at the screen. 'That's a heart-beat, isn't it?'

'Yeah, it is a heart-beat,' he said.

I started crying. 'Can you please get my sister?'

I had my hands over my face and I peered at the screen through my fingers, and there was Ella, like the perfect textbook embryo. She had her little fists raised as if she were waving. It turned out the bleeding was caused by poor attachment of the placenta.

I was still worried that the pregnancy wouldn't last. Someone suggested meditating and asking the baby if it wanted to stay. I thought that sounded a bit ridiculous, but if it would give me peace of mind I was willing to give it a go. So I sat down and I started listing all my thoughts and concerns. I said to the baby, 'Look, I know I am neurotic and obviously I am crazy because I'm talking to myself in a room on my own, but if you stay I promise I will try to be a good mum. You just tell me anything you need me to do, you just tell me.' I kept repeating these thoughts, and that was the first time I got any real response. And she said, 'If you'd let me get a word in!' I opened my eyes in shock. *Oh my God, she spoke to me!*

And we continued communicating right throughout my pregnancy.

I would ask, 'Are you there? Are you still there?'

'Yes,' she would always respond. I had a strong feeling that she was constantly with me. Sometimes I found myself almost watching my Ps and Qs because I wanted to do right by her. By the time Ella was born I had the utmost respect for her. I actually found it quite challenging seeing her as a baby.

The bleeding continued, although it slowed down, and I took iron tablets to combat anaemia. I was nervous a lot of the time, but I had horrendous morning sickness — which at least told me I was still pregnant. Despite all that I enjoyed my pregnancy, because at long last I was carrying a child. At nineteen to twenty weeks the bleeding stopped, and I went back to exercising and felt much better.

I started nesting the weekend before she came. I did a four-hour cleanout of the wardrobe on the Saturday, and cooked all Sunday. On the Sunday night we went to the movies to see *Love Actually*. Steve had gold-class cinema tickets that were about to expire. I was fine at the beginning of the movie, but when it ended I couldn't get out of the chair. I couldn't stand straight, and it was really uncomfortable to walk. I thought, *Oh, great, my pelvic floor's shot!* But I didn't complain. I'd been through months of bleeding and thought this was just the next thing to deal with. In fact, Ella's head had engaged, which was why it was so hard to walk. But that didn't occur to me because she wasn't due for another two-and-a-half months.

The next day, with great difficulty, I went to work as usual from 8:00 till 4:00 pm at my temp secretarial job. My belly looked really big, and all day the guys in the office were joking, 'If you need a ride to the hospital, I have a pretty fast car, ha, ha, ha.'

Even though I was quite fit, I was exhausted by the end of the day and was looking forward to getting into bed and watching *Wife Swap* on TV. I washed and dried my hair and, just as I was pulling the covers back and getting into bed, I felt this huge gush of liquid between my legs.

'Oh no!'

'What's wrong?' Steve said.

'I think I've wet my pants.'

I was angry with myself, thinking I hadn't done enough pelvic-floor exercises, and would now have to cope with incontinence. I went to the toilet, but the constant trickling just didn't stop. It never occurred to me my waters were breaking because I was only 30 weeks pregnant.

Steve called out to me, 'Are you okay in there?'

'Fine, I am just wetting myself,' I said. 'Can you please get me a new pair of pyjamas?'

I put a sanitary pad on and sat on the end of the bed, thinking it seemed a little strange. We sat looking at each other and I said, 'Maybe you should get the phone.' Just as Steve stood up, a huge gush of water rushed out of me onto the floor. That's when I got scared. 'Quick, get the phone!'

I started talking to Ella out loud.

'This is all up to you. If you need to come now, you come now. But I would really prefer it if you didn't.'

Stupidly, I had been saying to her from about 28 weeks, 'You come out any time you like after now, as long as you are healthy.' I wonder now if she took me literally.

Then I stood up and felt the jolt of her head engaging in my pelvis.

We got to the hospital at about 10:00 pm, and my doctor was still there from his day shift. Because I wasn't due for another ten weeks, Steve and I hadn't done any birth classes yet or even had an introduction to the hospital. We had no idea where anything was. We'd never even seen the delivery ward.

When they wheeled me in, my doctor said, 'What are you doing here?'

'I think I wet my pants.'

'I think you probably did, too, but let's check,' he said before examining me. 'No, your waters have broken. We'll admit you, and you'll have to stay in for the next seven weeks at least. If she hasn't come by then we'll induce you at 37 weeks. You are going to have to start on steroids to strengthen her lungs and antibiotics to prevent infection.'

'Why do you think she's coming so early?' I asked.

'The most likely cause is a weakening of the amniotic wall from an infection.'

He then gave me a shot of steroids to strengthen Ella's lungs. I only got one of those, although the plan was to have several. They also put me on a drip, and I had to have antibiotics every six hours. As it turned out, I only had time for two of those as well. I started having very mild contractions, but didn't recognise them as such. I assumed I'd be in hospital for seven weeks before the baby came. So when I started having pain, I put it down to stress, nerves, and tiredness. Steve said the contractions were coming every five minutes, but they were so mild that the midwives told him to go home to sleep. I was trying to be really quiet because there was some bimbo in the bed next to me, carrying on because she couldn't sleep with other people in her room. As the contractions got stronger, I lay on my side with my face up against the cold bed-rails trying to breathe quietly so I wouldn't disturb her.

To his credit, my doctor told me as much as he could about the realities of having a ten-week-premature baby. 'In the morning, we'll wheel you around to look at the Neonatal Intensive Care Unit or NICU. If she comes early, that's where she'll stay for a while. She will be small and will need support.'

But everyone assumed we'd have at least a few days up our sleeves, because you can break your waters and not go into labour immediately. The plan was to try and hold off delivery for as long as possible. But Ella had other plans; she just kept coming. It was about 1:00 am, and my doctor was still there finishing paperwork. He gave me an ultrasound and said, 'She looks like she's ready. You can see she's slowly pushing.' I could see Ella bouncing on my cervix, which I thought was a pretty cool thing to see. A nurse put the phone on my chest. 'We've dialled your number: tell your husband as calmly as possible to come back in and drive slowly.' Steve had only been asleep for ten minutes so he didn't take long to answer the phone. I told him, 'They reckon you should come back in because she's going to come tonight.'

I laboured sitting up and on my back because I couldn't move around much with the baby monitor strapped to me. The only painkiller I had was gas for about twenty minutes. I wanted to see how far I could go on my own using visualisation techniques I'd read in a book called *Mind Over Labour*. I had bought it two days before and had taken it to work on the Monday to read in my lunch break, and had only read about fourteen pages. During labour I threw the book at Steve and said, 'Just read it to me.' He has a very soothing voice, and for several hours I visualised going down into the troughs of a wave and then riding back up again. It gave me something to focus on.

The first few hours were slow and steady, but then it was bang-bang-bang. My cervix dilated really fast. It went from 3 cm to eight centimetres in about 40 minutes; it was really full-on at the end. Ella arrived at ten minutes past eight, that morning.

We only had a few seconds together because they had to get her under the lamp quickly before she got cold. I remember saying to her, 'Hello!' Finally I could put a face to her and make the connection between the baby and the sense of her I'd had for so many months. My first fear was there might be something wrong with her, not only because she was early, but because I was convinced I was carrying some sort of alien child. I didn't expect her to look as perfect as she did. When I first saw her she looked as if her fingers were fused together, but it must have been from the stickiness of the birth. She was quite purple and bruised for weeks where my hips had squeezed her during labour. A couple of times her heart almost stopped. In my haze, I remember seeing the staff in their white surgical gowns and caps on standby for an emergency caesarean. There were a lot of people rushing in and out of the room with clipboards and equipment.

But in the end Ella came out safely and breathing on her own. She looked so serene. She had really good facial expressions right from the start. I didn't see her for very long before they took her away. Steve went with her and I was stuck in a room that looked like a bombsite, and I didn't see her for the next three hours. Finally I got sick of waiting, so I walked around and all the nurses said, 'Hey, what are you doing? Get a wheelchair!' No one could believe I had walked all the way around to her ward. I ignored everyone and just sat by Ella's side. She had splints on her arms and a drip attached and tubes everywhere. There was a CPAP at her nose to help her breathe. She was so tiny, just 1.5 kilograms and 42 centimetres long, which is quite long for a baby that premature.

When she was inside me I could feel her, up the top left and bottom right, as if she was doing handstands. She used to do the same thing in her humidicrib. She would stick her fingers right down the edge between the glass and the bedding. The nurses hadn't seen that before, and thought it was funny. Usually when babies are that tiny they can't reach very far, let alone have the strength to even try. Ella was inquisitive and defiant from day one, and constantly tugging on her gastric tube. Unfortunately, she was always ripping things off her face. It must have been so uncomfortable. I can't blame her.

In the NICU the babies are put into one of seven bays

depending on the level of care they need. Bay 1 is the stage before you go home, and Bay 7 is for the most intensive care. Ella started in Bay 7 and moved up to Bay 3 after a few hours, which was outstanding for a baby of her gestation. She was taken off the CPAP because she had really strong lungs. Everyone was surprised and happy at her progress. I stayed in hospital with her for the first week and then had to go home, which was really hard. I remember thinking, *If we get hit by a car right now, there's someone we're responsible for other than the dog and the cat.* We hated leaving Ella in other people's care. I trusted most of the nurses, but not all of them. I put my foot down and said to one particular nurse I didn't want her looking after my baby ever again because I thought she wasn't gentle enough.

It was a strange feeling because finally we had a baby, but there was still no car seat in the back or pram in the boot. The nursery was all set up, but the cot was empty.

Meanwhile, Steve went back to work after a couple of weeks, which he didn't mind because it kept him distracted. I got into a routine and would be in the hospital every morning at 8:00 after doing my first milk-express at home. I would pack a bag with lunch, my expressing machine, magazines, and journals. I sat and wrote a journal every day to Ella about her. I would change her nappy whenever I was there. Some mothers let the nurses do everything, but I liked to be hands-on. I couldn't wait to get her out of that bloody plastic box.

I would come home at night, and Steve and I would throw down some dinner and both go back in for a few hours. I was absolutely exhausted. In hindsight, I don't know how I did it.

Some of the nurses would say to me, 'Go out and watch a movie, go shopping, it's not going to be long before you won't have this time to yourself.' I thought, *Are you insane? How could I go window-shopping? Get out of my way, I've got a baby to look at!*

I didn't want to do anything else other than just be there with Ella.

After two weeks, Ella's progress began to flag. She turned slightly blue because of poor blood saturation. One day they wheeled in a big machine and did an echogram of her heart. The cardiologist, whose name, ironically, was Dr Hope, had the unfortunate job of delivering the bad news.

'There is something wrong with Ella's heart,' she said. 'It is quite serious in a baby her size.'

I felt like my own heart was falling out.

Ella had truncus arteriosus, which meant the bottom two ventricles shared an aorta instead of having two separate ones. There was a hole in the bottom two chambers of her heart.

Ella's only hope was an operation, which had never been performed successfully on a baby that small. They liked babies to be at least double Ella's weight before surgery.

'Were you expecting this?' Dr Hope said. 'Does anyone in your family have heart problems?'

We couldn't think of anyone apart from Grandma, who died of a heart attack at 83 after a long innings of drinking and smoking. But as far as we knew there was no family history of congenital heart disease.

The doctors did what they call FISH, or Fluorescent In Situ Hybridisation, to determine whether it was a congenital problem or just a chance thing. It turned out Ella had no evidence of the chromosomal translocation Steve had. Instead, she had a congenital heart defect which, by a freak of nature, occurred on chromosome 22 — the same one Steve has a problem with.

To do the tests, they had to take a lot of blood from Ella, which was really upsetting, for her and me. She had bruises all over her, and her heels were permanently dark purple. Every morning there would be more pinpricks all over her chest, the back of her hands, everywhere. One day I shouted at the pathologists, 'Shit! What are you trying to do, drain her or something?'

Poor Ella screamed during those tests. I tried to calm her by singing to her. When she was in the womb I often sang, 'Dream a Little Dream'. The lyrics are, *'Nighty-night and kiss me/ just hold me tight and tell me you'll miss me/ while I'm alone and blue as can be/ dream a little dream of me.'*

It was sometimes the only thing that stopped her crying.

When Ella was in intensive care, I did a lot of fighting on her behalf. I was constantly getting out of my chair and standing eye to eye with any number of medical staff.

One day I was marched around to one of the registrars, who insisted I wasn't allowed to hold Ella anymore and must limit touching to the bare minimum. She claimed it upset Ella medically and made her heart work harder. But it was the way she said it that was so awful. She literally wagged her finger at me and said, 'No touching the baby. She can't be held.'

I was furious and said, 'Who are you? When did you come on duty? What is your name? Do you know who I am? Do you even know my baby's name?' Her bedside manner was absolutely abhorrent.

But the nurses would say, 'Touch is so important. She really needs to know you are there; she will grow strong and healthy faster if you are hands-on,' which is in line with my way of thinking. I truly believe it was that approach that helped my milk come in so fast and readily, which often doesn't happen with premature babies. It can be difficult to bond when your baby is in a special-care nursery and you aren't able to hold them all the time. I looked at photos of Ella to help me express milk at home. Normally when you have the baby at your breast, that connection makes the milk come in.

I was determined to fight for what I thought was best for Ella, and finally found a doctor who agreed with me and gave me permission to touch her.

A couple of times, I tried something they call kangaroo care, where you put the baby inside your top to have skin-to-skin contact. But poor Ella didn't really like being vertical and struggled a lot, and her sharp little fingernails scratched between my breasts.

But what was sometimes more disconcerting was that, even after Ella was born, I was still able to communicate telepathically with her when I was back home and she was in hospital. That troubled me because I thought, *You should be you, and I should be me, and we should be separate.* She was still with me too much rather than in her own body. It made me think perhaps she wasn't committed to this life.

Although Ella was slowly gaining weight, she got sicker because of the extra pressure it put on her heart. But she needed to get to 2.2 kilograms before the surgeons would do the heart operation. They moved her down to Bay 5 for more intensive care, and by that same night she was in Bay 7 again, right where she'd started with the CPAP on her nose to help her breathe. Whenever she was awake she would cry, and nothing — not even singing — would calm her. She got to 2.0 kilograms, so they moved her out into a little bed-warmer because she was holding her own temperature. That was the same day we took her last photos. She looked very sick because she'd also picked up a tummy bug. I wanted to know how she could possibly develop gastro when she was only on breast milk, but the nurses said she could have picked it up from a dirty milk bottle or someone not washing his or her hands.

By chance, that morning, Steve decided to skip work and come in to see her. He hadn't done that before. Just before we left home, one of the registrars called us.

'Ella is really sick,' she said.

'Sure, we know that.' I didn't take much notice at first because the registrar was new.

'She appears to have some sort of viral infection, so we're going to give her a general antibiotic which will hopefully cover everything. We don't know what it is yet. Are you coming in?'

'Yes, we're about to leave.'

When we got in Ella looked deathly pale. You almost couldn't discern her mouth from her cheeks or eyes; she was so pasty. She was out of it. Usually she was very interactive and always followed my voice. If I were across the other side of the room she would turn her head and look at me. She had been lots of fun for the nurses, too. People would just come and stand and look at her. This day, though, she was really listless, not even opening her eyes. Steve and I stayed by her side, holding her hand.

Then they did something scary; they wanted to put a tube right down her throat, to help her breathe, but they had to sedate her first.

The nurse gave her a needle and said, 'This will knock her out,' but when she flicked her arm Ella punched out at her.

'Ooh, she shouldn't do that; she shouldn't be able to move,' the nurse said.

Yeah, good on you, I thought. Ella was really strong; she was a little fighter. She had her fists up from day one. I was concerned she was an angry little baby, but people said, 'Look what she has to deal with. Wouldn't you be?'

The staff didn't want us to be there while they inserted the tube so, for the first time since Ella's birth, Steve and I went out for lunch. It was the first day I had left the hospital grounds and I said to the nurse, 'I've got my mobile with me; please ring if anything happens,' and for some reason the bloody thing didn't ring. When we got back, Janet, the head nurse, came running up to us, clearly flustered and upset.

'Oh, Kirrily, you're back,' she said. 'Ella's very sick. Where is your husband?'

'He's in the toilet.'

She went into the men's to get Steve while holding me up because she thought I might faint.

'We've been paging you for ages around the hospital.'

'We haven't been here,' I said.

I honestly thought she was over-reacting until we rounded the corner and saw the screen up around Ella.

'Janet, this isn't happening,' I said.

'I know, love, I know,' she said, 'I will get you a drink and a chair, but it is not looking good. You are going to have to be really strong.'

They were doing CPR on Ella, and every time they stopped, her heart stopped. They were keeping her alive until we got there. Steve crumbled; his face instantly filled with tears. I was stone cold. I couldn't cry or feel or even talk. There were people crowded all around her and everyone was just so sad. There were nurses crying behind the screen, and even the nurse doing the compression was visibly upset. I could tell Ella had already gone. I whispered in her ear, 'Please don't do this.' Ella had a tube down her throat and was completely expressionless. I wanted to say, 'Just stop. Give up.' Steve was sobbing and sobbing, but somehow I was calm.

The paediatrician said, 'We want to try one more thing. We have taken an X-ray of her lungs. Depending on what we find, we may be able to treat it. We are just waiting for the results.' Another few minutes passed and I thought, *I have to talk to Ella.* I put my hand on her little foot and closed my eyes and spoke to her silently, 'Okay, I know why you came and I know there are lessons here for me, but can't I learn them while you are here? Just tell me what I need to do.' And it was the first time she had spoken to me for ages. She said, 'Let go.' It was so strong. I said aloud, 'Okay.' I opened my eyes and the doctor put his hand gently on my shoulder and said, 'We have to stop now.'

They stopped the compressions and wrapped her up. The poor nurse who wrapped her was on her first day back from annual leave. She handed Ella to Steve, and she died in his arms.

'Do you want to bathe her?' one of the nurses asked me. 'It's very good for you. It's healthy.'

'I don't need to bathe her,' I said. 'I don't want my last memory of my daughter to be bathing and dressing her dead body. I don't want that image in my head.'

So they let us stay with her for as long as we wanted, in the room full of pinging machines and tiny babies in plastic boxes. We know the place that kept her alive for long enough for us to know her ultimately killed her. She got to 2.0 kg, 200 grams under the weight she needed to be for the heart operation. But we don't believe she would have survived that. At least we were with her, holding her and talking to her as she went.

After that, I didn't hear from Ella for a long time and only spoke to her in dreams. I had one dream where she was in the humidicrib and we were in a hospital in London, just the two of us. I was sitting next to her, and she turned and looked at me with a massive smile on her face. There have been other things as well. When I know they are not my thoughts, I know it's Ella talking to me.

For weeks after she died, I imagined I still felt her little nails on my chest. That was really hard, and I would often stick my forearm down between my boobs to put pressure where she had been, to comfort me.

Despite not having Ella, we were parents now. She taught us we are capable of having a baby. I know now what I am like as a mother, and it's how I always thought I'd be. She has taught me compassion and not to take things for granted. Sweating the small stuff is something I've always done; but, since having Ella, Steve and I have changed so much. He has allowed the softer side of him to open up, but he's not a pushover. Neither of us are. He is not as unforgiving of himself and others anymore. As a couple we've grown, too. We've been together for so long and from such a young age. We started our relationship on a childlike footing, but we have come through counselling, miscarriages, and our baby's death to make us grow up.

I had another pregnancy the following May, which I lost. We found out I was pregnant just before our fifth wedding anniversary. We went to the Sofitel in Melbourne, where we'd had our wedding night. It was a beautiful room, with a complimentary bottle of champagne. Steve doesn't drink, so I had the whole thing to myself. I'd started to miscarry five days before. I sat on the bed looking at the view thinking, *She should be here and I shouldn't be bleeding now, and all of this shouldn't be happening.*

Despite the sombre mood, we had a good night reconnecting. I am not surprised a lot of people don't last the first year of marriage. It has been hard for me not to resent Steve. Then I think, *Parenthood is his dream as well, and he wants it equally as much as I do.* We say to ourselves, 'This is not going to beat us.' We started out with beginner's enthusiasm, but over the years the number of people we know who've had kids easily is sometimes too much to cope with.

Six months later, I fell pregnant again and also miscarried. I didn't feel pregnant and had no bleeding, which was unusual. At the first ultrasound the heartbeat was so slow that they thought it was mine. The doctor said, 'It's not a matter of "if" you miscarry, it's "when." ' It's three days before Christmas, and I would rather catch it and save you the pain of miscarrying on Christmas Day.'

They gave me a D&C and tested the embryo. It was a girl, and she had too much of one chromosome and not enough of another. When I woke up, the doctor was there. 'It all went well. We got everything,' he said. 'How do you feel?'

'I never want to meet you like this again,' I said. 'I'm going to do IVF.'

We had formed a very close bond. He and his wife had lost their first baby, too. I didn't find that out until he and his midwife shouted me lunch about four days after Ella's death. He's now on a mission to get us a family. He had been gently guiding us towards IVF because it means they can genetically screen the blastocysts before transferring them to my uterus.

Finally, after six miscarriages and Ella's death, we committed to IVF. That was a scary jump for me. I'd always said I was never going to do IVF, and that if it didn't happen naturally it wasn't meant to happen. But I think you get to a point where you don't want to die wondering 'what if?' For some people that point might come six months after trying or six years down the track.

We did the first round of IVF in June, just before my 30th birthday. I remember thinking I'd either be drinking lots of champagne on my birthday to drown my sorrows or celebrating with apple juice. I found the IVF drugs shocking. By the third day I was a screaming banshee. I did all the injections myself except for the big trigger needle. I'd read the instructions that said, 'Jab the skin at a 45-degree angle in a dart-like motion', and I said, 'No way! I'm not doing that!'

Steve said he'd do it, and practised on an old rockmelon. Then he pulled out this massive syringe and jabbed me, which really hurt.

'Good. Is it in?' I said.

'No, it bounced off. I'm sorry,' he said.

'Oh, great. Now you're blunting it.'

Because he was so scared of hurting me it ended up being even more painful.

We only got six eggs at the pick-up, which is abysmally low. You want at least ten good-quality ones for the pre-implantation genetic diagnosis, or PGD. They injected the eggs with Steve's sperm and only three fertilised. I was sick to the stomach with nerves waiting for the results. The egg collection was Friday, they fertilised them that afternoon, and we found out Saturday that half had fertilised. Then we had to wait 'til Monday to see how many had survived the weekend and were strong enough for the biopsy. They analysed all three and then we had to wait another day for the results. The lab called to say only one of the three was normal. They let the other two succumb, and we were hopeful the normal one would work.

Fortunately the embryo recovered from the biopsy after they'd extracted two cells from it. Apparently we have quite strong embryos. Even the abnormal ones look perfectly healthy. They put the good one back in me, and we were cautiously optimistic. But we've had too much tragedy in our lives to be that naïve. Anyway, I got my period four days before the pregnancy test was due. We didn't fall in a heap straight away, but we did later. When we had pulled ourselves up again, we said, 'Right, we'll try one more time before the end of the year.'

We did another round of IVF in August. We were excited about this cycle, but didn't tell anyone except for some friends I'd made on an internet support group.

This time, at egg harvest we ended up with seven eggs. Four fertilised, but one of those didn't continue dividing overnight.

Once again the lab tested three embryos, and once again one was normal and two were abnormal — the same odds the geneticist gave us back when we discovered Steve's condition. The lab also tested Steve's sperm again and found he had a very good sperm-count with a high volume of straight swimmers; but when they analysed the DNA, they found only 26 per cent of the sperm were normal.

The good embryo was transferred, and it implanted. I knew I was pregnant because I had done a couple of urine tests at home. We were quietly confident this was going to work because it was a normal embryo and we knew I was capable of carrying a baby. I was also having blood-thinning injections to counter an elevated level of antibodies in my system, which might fight the embryo as a foreign body and create blood clots around it. Those injections were far worse than any trigger injection. It was like sticking a darning needle in my stomach. I also took aspirin every day to try and keep the blood thin. We felt we had every base covered.

One morning I woke suddenly at quarter past three and felt Ella really close. *That's nice*, I thought and then, *Hang on a minute. Why are you here?* I became aware of a sensation in my tummy. It wasn't sore or cramping, just a pulsing motion, coming and going, and I thought, *It's leaving*. Ella felt so close, and I knew she was there for comfort. I didn't want to get up and go to the toilet because I knew what I would find. Instead, I lay there with my heart pounding. Eventually, I got up and, even in the dark, I could see two very big spots on my panty liner. Five hours later it was in full flow. I had a pregnancy test at the clinic later that day which was positive but showed a very low reading of HCG, the pregnancy hormone.

By Wednesday night I was rendered horizontal on the couch in pain and having cold sweats. I knew my body was going through another miscarriage.

So that's where we are up to now. After two rounds of IVF we've decided to try on our own again without medical intervention. The odds are the same, and the IVF and PGD combination is very expensive; two cycles have cost us around $10,000. We're also concerned about the toll of the genetic testing on the embryos.

People ask us how we keep going. I think it's because I had that connection with Ella, that I know I can have it again. Sometimes you need medical assistance to make the physical happen. I can see why people do IVF. It gives you something to focus on and look forward to rather than being in a void. We are glad we gave it a shot. At the end of the day, if we have no more children, at least I won't look at Steve and say, 'Why didn't we try PGD?' I'm not saying we won't go back to it; but, for now, we want to try on our own.

During this last cycle Steve said to me for the first time, 'I hate the thought, but maybe we need to think about donor sperm.' The difficult thing is we've seen what his sperm and my egg can produce in Ella. She didn't have his chromosomal condition. The IVF also showed us we could produce normal embryos. Consequently, we've decided to keep trying with Steve's sperm.

'Just cope through the next few. I am sure we will get there,' he said. We have a belief that if we can get through the miscarriages together, eventually we'll be blessed with a normal baby with a normal heart and born at term.

Some people say, 'Keep your eye on the prize.' I'd take that one step further and tell myself to make a connection with the bigger picture. It's not just a baby we're trying for; it's a baby who's going to be a child, who's going to be a teenager and an adult who will hopefully see us to the end of our days. Maybe this person will go on to make a difference in politics or saving the whales in 30 years' time. Perhaps we have to endure ten years of miscarriages for that end result. When I look at it like that, it snaps me out of my own ego stuff.

Postscript

After eight miscarriages and Ella's death, Kirrily finally gave birth in June 2006 to a healthy baby girl that she and Steve named Lauryn. Lauryn was conceived naturally, and Kirrily gave birth to her at 39 weeks. The couple are not contemplating any more children, at least for the time being:

> We still weave Ella into our daily lives. At Lauryn's bedtime we pass a big picture of Ella asleep in her humidicrib. I point and say to her little sister, 'Baby Ella is asleep. Time for your sleep, too,' and now Lauryn looks for the sleeping baby before bed. I love that this special ritual has become one of her sleep cues. It's important to me as much as it is comforting for Lauryn that she knows bed and tucking-in comes next. It breaks my heart that they will never meet, except perhaps in her dreams.

CHAPTER 2

Ugly Brown Eggs
–Ada's story

Ada has been a member of the essentialbaby on-line community for almost three years — which, at times, has been her lifeline. She and the other IVF veterans on the support chat-room call themselves the 'tough nuts' because they've been through so many unsuccessful cycles. Despite her usual stoicism and positive attitude, Ada lapsed into depression when she and her husband were told they would need donor sperm and eggs.

> I was devastated about needing donor eggs, but I didn't know what devastation was until we discovered we'd also need donor sperm. That was probably the single worst part for me, coming to terms with the fact we wouldn't have our biological child.

I'm one of six children, and my mother was one of sixteen. My mum's Russian and my dad was Ukrainian. I always expected I would have children. You really don't think you're going to have problems when you come from such a big family. In fact, having kids was pretty much the only expectation I did have. I certainly wasn't the type of girl who thought, *I'm going to have a big white wedding, a wonderful husband, and a house with a picket fence.* I'd seen people breaking up, including my own parents, and thought I'd probably have a few relationships and, when I'm about 27, if I'm on my own, I'll just pick up a guy in a nightclub and get pregnant. That was the plan.

Of course, it all turned out very differently.

Having a guy around wasn't an important issue for me. I'm not terribly romantic. The funny thing is, Julian is the gushy, romantic, starry-eyed type.

Julian and I have been together for nineteen years, but only married recently. We met when we were nineteen and twenty. I'd just started going out with a boy called Tim; it wasn't really serious, we were just having fun. I was at his place one day and he said, 'My best friend's turning up. I can't wait for you to meet him.' Julian walked in and I thought, *Oh my God, what a spunk!* Tim asked Julian later, 'So, what did you think of her?'

'She's the sort of girl I could easily fall in love with,' he said.

A few months later Tim dumped me and Julian said, 'You can cry on my shoulder!' We've been together ever since.

I just can't believe how lucky I am to have Julian. He's pretty much perfect. If I could change anything about him, I'd probably make him an inch or two taller. But other than that, I wouldn't change a thing.

At first we shared a house for a while and then travelled together. When we bought our own home we felt like a couple of kids playing grown-ups.

We had a little pregnancy scare in the first couple of months we were living together — my period was a few days late. We weren't using the pill, but decided if I fell pregnant that it would be okay, so we continued using the withdrawal method. We weren't going to plan for children until we were financially secure, but we definitely wanted them.

At the time, my Uncle Paul, who's in the building trade, said to us, 'Buy the biggest, most expensive house you can afford. Stretch yourself to the absolute limit; you'll struggle for the first couple of years, but wages and real estate values will go up and, within a couple of years, you'll find it won't seem so difficult. By the time you're finished you'll have a really big house and a good nest egg.' We thought that sounded like a good plan, so we took his advice.

Julian was working in IT and making good money, and I had a job in the public service. I worked my way up, and now we're both in IT. By the time we got the deposit, we were both being paid reasonably well, and the mortgage, as my uncle had predicted, had become more manageable. When I was 33, I got a redundancy payout from work, which took care of the mortgage completely. Finally we were debt free, and we said, 'Right, let's start a family now.' But then I got a new job, and — this has to be right up there with one of the stupidest things I've ever done in my life — I thought, *It's unfair to my boss to get pregnant in the first couple of months. I'll wait a year and then start trying.* How dumb is that?

Consequently, I was 34 when we started trying. We'd been together fourteen years. At first it was pretty funny, because we were open with our friends and said, 'Okay, we're trying for a baby, so nobody phone us at night because we'll be busy!'

We started charting my ovulation, and when baby-making night came around, we lay there for a few minutes and then Julian started laughing, 'This is crazy, I can't do this. Where's the romance? I can't just perform on demand!'

'Well, we're never going to get there if you don't get to work.'

Then I remembered a girlfriend told me sometimes she and her partner would pretend he was a male gigolo and she was paying him to perform. So we tried that, which was fun, and everything worked just fine.

We tried to conceive for almost a year, and nothing happened. Something in the back of my mind told me maybe there was a problem, but Julian was very logical, 'Everyone says you have to try for a year before you seek assistance,' he said. We saw a

GP after eight months, and she also said we should really try for twelve months. However, I was as keen as mustard and said, 'No, I want to see an obstetrician now.' But Julian was adamant we wait out the twelve months.

My instincts still told me there might be something wrong, although I didn't think it was serious. I thought it would be a simple fix with a pill or something.

Our first obstetrician was also really upbeat and confident. 'You'll be fine. We've done all the tests. Everything's clear. We'll get you pregnant in no time.'

For the next twelve months we tried the fertility drug Clomid and intra-uterine insemination, where they trigger your ovulation and then inject the sperm into the uterus. I was convinced it was going to work. However, after the tenth IUI I thought, *I'm getting bored with this. Let's move on to the big guns and do IVF.* At that point I was impatient rather than desperate.

We only told a few people what we were doing. We didn't tell my mother; she still doesn't know. She's hypersensitive and prone to theatrics. As far as she's concerned, we live happily together without children by choice. Occasionally she says, 'I'm proud of you for choosing to be child-free.'

My dad never knew. He died two years ago, on the day I did my last IUI. It was a terrible shock. He was seventy-five. It just came out of the blue. He'd been moving railway sleepers around the backyard the week before, and the next thing you know, he was dead.

On the morning I was having my last IUI, I got a phone call to hurry to the hospital because Dad's pneumonia had got worse and there were complications. He was sedated and on life support. It should have been obvious to us he was going to die, but we didn't think he was that sick. Looking back now, I was in denial. I first realised how serious it was when the doctors said, 'We're turning the machines off now.' I should have asked more questions about what was going on. I felt sick for not being stronger. My sisters and I still question whether more could have been done. He was in hospital for a week before he died, and the specialists talked about doing certain tests; but later, when we asked if they'd been done, they said, 'Oh no, he wasn't well enough for us to do that.' In the end, it wasn't the pneumonia that killed him; it was a blockage in his artery. I still wonder, if they'd done those tests, could they have saved him?

I remember thinking, *Dad's up there now and he'll help us with this IUI cycle. We'll have a little boy and name him Tom after my dad*. But it didn't work, so I thought, *Maybe Dad needs to settle in and find his way around first*. Knowing my dad, he probably thinks we're better off without kids anyway. He thinks he's doing us a favour!

Julian was ready to move on to IVF before me. He had to wait for me to catch up. I was lucky not to have any adverse reactions to the drugs, although injecting was initially difficult. The very first time I drove to a friend's place and got him to inject me because neither Julian nor I could cope with it.

Julian's the sort of person who, if he has to have an injection, has to lie down or else he'll faint. I said to him, 'If I'm going to go through IVF, then injecting me is the least you can do. So brace yourself and deal with it.'

I knew it was going to be a drama the first morning we had to inject on our own, so I intentionally didn't remind him until he had to race for the train, then I said, 'Quick! Hurry up, get that needle ready.' He didn't have time to worry. It was much easier after that.

We only got four eggs at the first pick-up. Two fertilised, and they put both back in me. I actually made it to the day of the blood test without my period showing up. I was completely confident I was pregnant. I didn't know then about the effects of the pessaries and other drugs. As far as I knew, if your period doesn't come, you're pregnant. When I did the blood test at the clinic and the result was negative, I swear I nearly fell over with shock. Julian and I stood in the hallway hugging and crying because we just couldn't believe it could be that cruel.

I had to wait for a month before we tried again because my transfer was really difficult; I bled during the procedure and it was awful. My uterus is fishhook shaped rather than the usual gentle dish shape, which means it's difficult to get the catheter inside. Apparently the ease of transfer is a big part of the success rate. The second time around, my obstetrician suggested we do GIFT — gamete intrafallopian transfer. They take the egg and the sperm and squirt them together into the fallopian tubes and let the fertilisation happen there. The result was, again, negative. The third time we tried another method called ZIFT — zygote intrafallopian tube transfer — where they fertilise the egg and sperm in a dish, and then put them straight back into you.

We also started another couple of rounds, which were cancelled half-way through because I developed a cyst on my ovaries. When that happens they can't stimulate your ovaries because the cyst absorbs all the stimulants, and grows bigger and bigger. It happened twice, and in both cases I had to immediately stop the drugs and take the contraceptive pill for a month until the cyst quietened down and was absorbed by my body.

I can usually put a positive spin on most things. But I started to get depressed when the clinic said, 'Your eggs aren't good. You don't produce a lot, and there's a problem with the quality and the grade.' When I asked how they could tell, they said the proof was in the fertilisation rate. Usually, if there is a problem with the semen, it shows up in the sperm analysis. Julian's tests were all fine, therefore it was a fair assumption the problem lay with my eggs.

I'll never forget the day I rang the IVF lab to talk to the scientist about my egg quality. The technician said she was busy, but then said, 'Let me have a look at what she's written on your file,' and he read out, 'Ugly brown eggs'. I felt like I'd been slapped in the face. I don't think people realise how much words can affect you, especially at such a vulnerable time.

That was the first time it occurred to me I might not get pregnant. Over time I became depressed and retreated into myself. I'd say to Julian, 'You know last week, when I was really quiet? Well, I was really down then. But the fact that I'm talking about it now means I'm feeling better.'

After the third attempt, the doctor called us in to his office. He didn't have on his usual happy face. Instead we got the serious 'you're going to need a donor' talk. I thought, *No way. I'm going for a second opinion.*

The second obstetrician I saw worked for a religious hospital and said they didn't deal with donor eggs. So I had to find a third obstetrician. I was too scared to go back to the first one. I felt bad for leaving him, like I'd cheated on him. I found another doctor, who I'd heard good things about. By that stage we'd come to accept that we'd have to use donor eggs.

My sister, who is ten years younger than me, offered her eggs. She was the perfect candidate, aged 27, and had already had a child. She agreed to donate almost immediately, before I'd told her the details.

'Well, hang on a second,' I said. 'Do you realise you could die? Do you realise you could hyper-stimulate on the drugs? Have you thought about the consequences for you and your family?' I was so negative, my other sister said, 'For God's sake, Ada, what are you trying to do? Talk her out of it?'

'I just want to make sure she knows what she's getting into!'

Fortunately, she still wanted to do it. Her husband also agreed, but on one condition — that we keep it a secret from the child. But we couldn't agree to that; it was something we felt strongly about. I don't think secrets are a good idea in a family, unless you're sure you can keep it, like I can with my mother.

My brother-in-law has issues about being adopted. He would have preferred not to know he was adopted. He was concerned about our child turning up on his doorstep one day and saying, 'Hi, I'm moving in with you guys. I'm not staying with those people.' I said, 'You know what? Your daughter could turn up on our doorstep one day, too, and say the same thing!'

After discussing it at length he was finally comfortable with the idea, and we all agreed to go ahead. My sister coped well with the drugs and produced sixteen eggs. I thought, *This is it — this is the magic mix!* But then something strange happened. The fertilisation rate was very lacklustre. Only eight of the sixteen fertilised which, for a 27-year-old who was proven fertile, wasn't fantastic. By the time we got to day two or three for the embryo transfer, only two were looking okay. I thought, *That's a bit weird. Her eggs are performing as badly as mine.*

The clinician was also suspicious, and ordered a new test called a SCSA or Sperm Chromatin Structure Assay. It shows any DNA damage to the sperm with a fluorescent marker, and enables them to calculate exactly what percentage is damaged. Julian had a reading of 29 per cent abnormal. I'd already Googled the results, and knew that anything from 30 per cent upward meant he was unlikely to conceive.

We're not really sure why Julian has a problem. Things like chemicals in the environment and lifestyle can cause male infertility. There are ways to reduce the damage over time with antioxidants or ejaculating more frequently. Ironically, one of the clinic's recommendations is for men to abstain from ejaculating a few days before the sperm sample is needed, which is actually one of the worst things for people with high DNA damage to do. Some men are better off ejaculating right up to the time they need the sample to get the freshest swimmers.

I was devastated about needing donor eggs, but I didn't know what devastation was until we discovered we'd also need donor sperm. That was probably the single worst part for me, coming to terms with the fact we wouldn't have our biological child.

Amazingly, Julian was fine about it. We were sitting in the doctor's office and he said, 'You're going to need a sperm donor,' and I heard Julian say, 'Donor? Okay, no problem.' I could feel myself starting to fuzz, like when I'm going to lose it. I held myself together in the surgery but, as we walked out, Julian said, 'Are you okay?' And I just lost it. He put me in a chair in the waiting room, and I bawled and bawled. It was horrible.

It was harder for me to give up the idea of having Julian's child than it was for me to give up the idea of using my eggs. Perhaps having my sister as our donor makes it easier, but I still have trouble accepting it.

My best friend's husband offered to be our sperm donor before we'd even asked. I found talking to him about the process surprisingly squirmy. Before you donate sperm you have to make sure your genitals are clean. How do you tell your best friend's husband, 'Make sure you wash your balls.'? It's much harder to talk about sperm donation because it involves orgasms and erections and it's sexual, whereas egg donation is quite clinical.

My best friend and her husband were very good about it. The day after he donated the first time, I got a message from him, saying, 'What! No flowers, no chocolate, we don't talk anymore! Where's the romance gone?' So I sent him a bunch of flowers at work saying, 'Thank you for yesterday, you were terrific.' Fortunately, Julian is not the macho type and was surprisingly good about the whole thing, too.

Anyway, when we decided to use donor sperm, we thought we'd go back to my eggs and see if that combination worked.

This time, three of my eggs fertilised, so we had them all put back in. I had a laparoscopy again so they could put the fertilised eggs in through my belly button. I was really annoyed with the surgeon because he cut the scar vertically instead of horizontally. All the other doctors had cut horizontally on the same very neat line. He just went and put a vertical line in — now my belly button looks like a hot cross bun.

I was far from convinced it was going to work. Julian and I had already decided to keep our options open, and we were ready to adopt. However, to adopt in Queensland you have to be married. I thought, *Well, I suppose I'd better organise this wedding if we're going to get anywhere.*

As I was on the phone, writing down the phone number for the Births, Deaths and Marriages Office, I thought I'd quickly ring the clinic and get my IVF results. When the nurse said, 'Positive,' I nearly screamed. I was at work and my colleagues were all around me, so I said, 'Look, I'll call you back.' I put the phone down and ran to the computer room and rang my best friend and started screaming. I was hysterical. She thought somebody had died. You wouldn't have thought I was trying to get pregnant. I was so shocked. And it was a really strong HCG reading, too.

Julian was in a conference all day, so I couldn't get directly through to him. I text messaged him '+ + +'. He was giving a lecture and had his phone on silent. As he was talking, he saw the phone vibrating on the table, and casually picked it up and saw three plus signs. He just kept talking as if nothing had happened; but in his head he was thinking, *Hell, we're having triplets. I'll have to get another mortgage!*

As it turned out, by the time we had an ultrasound, there was only one heartbeat. Still, I didn't let myself get too excited.

Then, at nine weeks I had a spot of bleeding, and the next morning I raced into the clinic. There was no heartbeat.

That was really, really hard. I was lying on the examination table and Julian was standing between two obstetricians. I put my hands over my ears and said, 'No, no, I don't want to hear this,' and then I realised the doctors were helping Julian off the floor. He was crumpled in a heap. I didn't know what to do. It was just horrible.

I remember thinking I had actually managed to get pregnant after five years. I held on to the fact that I'd finally done it and got to nine weeks. Goddammit, I almost got a quarter of the way there! That was much more than I'd previously achieved. That alone gave me hope.

I talked to a counsellor after the miscarriage, but didn't find it very useful. I have close friends who are good to talk to, and also my support buddies on-line. Talking to people who've been through it is very important for me. I can't talk to a counsellor, I feel silly; it doesn't work for me.

After the miscarriage, we decided to go back and try one more time with my 'ugly brown eggs'. But this time the procedure went really badly.

My obstetrician thought my cervix might have readjusted itself after the miscarriage and the D&C (dilatation and curettage). So, rather than doing a laparoscopy, where they feed the catheter through my navel under general anaesthetic, he'd try going up the cervix again while I was awake. He did a dry run the day before and all seemed okay. But on the day it was terrible. It took 30 excruciating minutes to get the catheter inside me. Because the ease of the transfer contributes to the success rate, I obviously wasn't terribly hopeful. I wasn't at all surprised when I got my period a few days later.

By this stage, I knew we would have to use my sister's eggs again with the donor sperm. And that's where we are now. We should start the next cycle in a couple of weeks.

My sister and I are taking hormones to synchronise our cycles. Some clinics do a frozen-embryo transfer when an egg donor is involved. They'll harvest the donor, fertilise the eggs, and then freeze the embryos until the recipient is ready. But other clinics, like ours, actually force the two women into sync. I'm happy about that because the success rate with fresh-embryo transfers is usually higher than with frozen embryos.

We're lucky my sister and I were only two days apart with our cycles this time. I ring her in the evening to remind her to take her nasal spray and say, 'Knock, knock. It's the drug pusher!'

My sister's excited and says, 'I can't wait to see you with a baby!' But underneath we're not overly hopeful. This will be our eighth attempt.

Among my group of friends there's a scary number who've been through various fertility problems. My best friend, the one whose husband is our donor, has had several miscarriages. My next closest friend, who's in her mid-thirties, had to have an emergency hysterectomy a few years ago. She had a fibroid the size of a football. She'd only just married and was just back from her honeymoon. They don't have children, and it's been really hard for them to accept.

When a friend falls pregnant I feel nothing but happiness for them, but when I hear about an acquaintance having a baby it hits me like a tonne of bricks.

We're lucky that Julian and I talk a lot. I'm not the type of person who's prone to chucking hissy fits or being irrational. If I'm upset, I'll say I'm upset. If I'm feeling irrational, I'll say, 'I'm feeling really cranky for no reason — so if I get snappy, I'm sorry in advance.'

Very rarely these days do I see myself holding my own baby in the future. Occasionally, when I'm in the right frame of mind, I can, but it's rare. It's almost like I'm going through the motions of IVF now. It's a job; it's what I do.

We're back to thinking seriously about adoption. Since I was a little girl, I've fantasised about rescuing the starving children I saw on TV, giving them a nice place to live, and taking care of them. That was until I found out how hard adoption is.

Here in Queensland, you've got to be married for at least three years before you can even apply, which is crazy because Julian and I have been together for nineteen years. We're planning to get married, but there is still a stubborn part of me that thinks, *Goddammit, I don't want to get married for this reason.* We resent being forced into this corner. We're not treating this wedding as something dreamy and wonderful; we're just going to the registry office to sign the paperwork and get it over and done with.

We're also thinking of moving to Canberra. I've done some digging around, and found out they process adoption applications much faster than in Queensland. It takes about fifteen to eighteen months in the ACT, whereas it takes about two-and-a-half years in most other states. Fortunately, I know of a few people from the essentialbaby on-line support group who live near Canberra. It's hard to think about leaving Brisbane because my friends are like sisters to me. I rely on them so much.

But once we have adopted a baby we'd move back to Queensland. We've chosen to adopt from China because they like you to have some sort of cultural link with the country. Although my mother's Russian, she was born in China and grew up there. So I think that's as good a link as any.

In every other state you call the department and say, 'I'd like to formally apply to adopt a child from China.' But in Queensland you have to wait until they open their doors for six weeks only to register expressions of interest. Then that's it for another three years. And now Queensland has closed its books indefinitely. So, if you've missed that window of opportunity, tough luck. They've got a stack of five or six hundred applications on their desk. They sort them into who they think are good and bad candidates: 'Okay, you have a Chinese person in your family: you go to the top of the pile. You don't have any cultural links, so you go to the bottom.'

I've heard of people who haven't been able to get through the process, so I know there's no guarantee. I wish adoption wasn't so hammering. I feel like I've been through enough torture.

My advice to people who are having difficulties conceiving is not to assume it's going to be all right. Get all your blood and sperm tests done immediately. Don't put it off for another month because every one of those months adds up. Don't keep hoping for a miracle. Look at me. I did ten IUIs, convinced I was just having a bad month. I look back and think I was an idiot. I've now done ten IUIs and eight IVFs in five years.

It's also very expensive. We've probably spent about $40,000 trying to have a baby. Ironically, thanks to Uncle Paul's advice, we have no mortgage. We have no expenses except IVF. It's also all-encompassing; it's become our hobby! I used to like renovating and fixing things up and I'm fairly crafty, but that's all gone by the by. Instead, we do IVF. It's hard to know what will fill the void if we stop. We've been doing IVF obsessively and compulsively for what seems like forever.

None of the doctors have ever said, 'Stop.' They leave it up to you. Usually, specialists take charge; but with IVF, you're only limited by your chequebook. Julian has also never said we should give up. He knows how incredibly important it is to me. There has never been anything I've wanted as badly as this. I don't think he'd ever try and stop me. If he did we'd be in for a few fights.

Through all this, I've been forced to grow up. Until my early thirties, I saw everything through rose-coloured glasses. Now I feel emotionally stripped bare, and I worry, *What if I have a child and I still don't feel better?*

I compare myself to battle-weary Vietnam veterans who are scarred because of the horrors they've endured. They don't really get better. What if that's me? What if I don't ever soften up? I'm like a crusty, old, toughened veteran in my khakis smoking a cigar and talking on-line to these fluffy young IVFers who are doing their first round. I'm one of the 'Tough Nuts'.

I have asked the girls on-line, 'How do you feel after you give up IVF? Do you feel scarred by it? Do you still look at people with a baby in their arms, who've got pregnant easily, and feel bitter? Will that feeling ever go away?' Some say it does, but I can read between the lines and some scars stay forever.

I'm not religious at all, but I have found myself bargaining with the gods and goddesses, 'If you're up there, please help me out.' At one stage I had quite a collection of offerings next to our bed; there was a rabbit, a fertility goddess, and a little bag stuffed with herbs. I used to wonder what the hell our cleaning lady thought.

A girlfriend of mine said recently, 'You've got a lovely house and great jobs, and you've got each other. Isn't that enough?'

'You know what? It's not enough. I thought it was, but it's not.'

This is the biggest and hardest thing I've ever had to deal with in my life, and that includes holding my father's hand while he died. As hard as that was, this is just so much harder. Nothing else comes close.

Postscript

Unfortunately, IVF number eight with my sister's eggs and donor sperm didn't work. Julian and I decided to throw caution to the wind and start the adoption process, which meant moving interstate. We moved over the border to New South Wales, and we commute 90 minutes to work in Brisbane. We are still relatively close to family and friends. God knows, we wouldn't have got through the past six years without them. We found a cute little house in Tweed Heads, and by December it was ours. We sent our adoption paperwork off the minute we got the keys.

Around the same time, a friend of a friend approached us out of the blue and offered to donate her eggs. I told her about the adoption and she said, 'Why not have two irons in the fire?' So, using her eggs, we started our ninth and final IVF cycle in January, and the transfer went ahead in early February. We really didn't have much hope for this cycle. It was a fairly half-hearted, last-ditch effort on our part. It was such a shock when the clinic rang and said my pregnancy test was positive. It was so strange, because mentally I had already moved on and was focusing all my attention on adopting from China. I almost resented being dragged back into the IVF world again. We were very sceptical after last year's miscarriage.

I had a rough time at about six weeks: I started bleeding badly, and was told we had a 50/50 chance of losing it due to a huge blood clot in my uterus. I was put on 40 milligrams of Valium a day, and told to bed-rest for six weeks.

At the same time we had to decide whether we were going to continue with the adoption process or not, because part of the deal is you must withdraw your application if you fall pregnant. We were about to spend another two thousand dollars on a two-day information seminar about adopting a Chinese child. Adoption ends up costing around $30,000 — that's about $10,000 for the department in Australia and $20,000 to the adoption agency in China. We didn't know what to do. It was a gamble. We decided to withdraw and bank on the pregnancy going ahead. Thankfully the scare passed and I'm still pregnant. At least we backed the right horse this time!

I'm now 30 weeks pregnant and feeling fantastic. I didn't realise how emotionally down I'd been until I got pregnant and the black mood lifted. Because it's been such a hard road, we put off telling people until I was sixteen weeks pregnant. We've told my mother, but we haven't told her about the donor eggs and sperm. According to the ultrasound, we're having a girl, which is funny because I was convinced we were having a boy. We're due in mid-October.

It's taken a global village to make this baby. The egg donor, Paula, is a single mum with a little boy. I was worried she'd contact me all the time to see how the pregnancy was going, and I would feel beholden to her. But I don't now; I really feel like this baby is ours. Paula has been fantastic. We work in the same building and have a catch-up and a coffee every couple of weeks, but she leaves it to me to initiate contact. She's an earth mother, and just wants us to be happy. I'm confident she doesn't feel as if the baby is hers. She's happy to be called Aunty Paula.

Warrick, our sperm donor, doesn't think of himself as dad, either. We keep reminding him he helped make this pregnancy possible. His wife and I are best friends, like sisters. We've known each other since we were eight years old. I'd like to make Paula and Warrick the baby's godparents.

The reality of this pregnancy has dawned on us slowly. I was in a pre-natal yoga class and I looked around at the pregnant women and thought, *My goodness, I am one of them!*

Several weeks later, Julian and I started our childbirth classes, and at one stage Julian turned grey and had to sit down. I said, 'What's wrong, are you going to faint?' and he said, 'It's really happening, isn't it? We're really going to be parents.' He's now getting into the swing of it, reading pregnancy and childbirth books, and constantly asks me if I can feel the baby kicking. He finds that re-assuring, but still worries something could go wrong.

For a while I worried I might not feel bonded to this baby because biologically it's not mine. But those fears have completely vanished. This is my baby. I'm hoping I can breastfeed, because so far my part in making this baby has been relatively passive, and breastfeeding will be my gift to her.

The funny thing is, after almost twenty years together, Julian and I were married in a registry office last October as part of the adoption process. We haven't got used to calling each other 'husband' and 'wife' yet. I wonder how we'll go when our child starts calling us 'Mummy' and 'Daddy'?

CHAPTER 3

Sometimes It Takes Three to Make a Baby
–Caroline's and Drew's stories

After yet another miscarriage, Caroline awoke in the middle of the night crying, 'If I can't have children, what will I do with all this love inside me?' The next day her husband bought her a labradoodle pup. But it was the gift from an egg donor which helped fill the void.

> It's been a huge emotional roller-coaster for all of us. It's taken three of us to make this dream come true.

Drew: I come from a broken family. My parents separated when I was three and I had no real concept of family, let alone marriage. Although I aspired to be married one day, I didn't have a strong vision of children and a family.

Caroline: I, on the other hand, had a very small but incredibly close-knit family. I have no cousins or grandparents, but my parents, sister, and I existed in a very supportive and loving unit. We love being together and have a lot of fun. Sadly, Dad died about three years ago. My parents had an excellent marriage. As a result, I've always had a very clear picture of how I wanted my own marriage to be, which was why I waited so long to find the right person. My parents also had children late; Mum was 36 and Dad was forty-six. That was unusual for those days, and so I never felt any pressure to have kids early. My parents never said, 'Ooh, when are you having children?' Instead they said, 'Live your life. Have a ball.' I always thought having children was something you did later. I had no idea how difficult it would be.

I was never a career woman as such, although I had some fantastic jobs and a hell of a lot of fun. For a while I was a fashion buyer and then started my own retail and wholesale businesses. I also lived overseas for a long time and enjoyed travelling. I never worried about the future.

By the time Drew and I got married I was 38 and he was a year older. We actually met for the first time when we were about four years old in our hometown of Melbourne. We even went to dancing classes together when we were twelve. Although we moved in similar circles for years, we didn't know each other that well. About fourteen years ago, by coincidence, Drew and I moved to Sydney at the same time. He'd been in London and I'd been living in Asia.

Seven years later our romance took off quite quickly. Then, not long before we married, I got pregnant and miscarried. We dealt with that relatively well, but it was the next miscarriage six months after we married that threw us for a six. It was around the time of Drew's 40th birthday. We were very excited about this pregnancy, and were booked in to have the CVS (chorionic villus sampling) test at twelve weeks to check for Down's syndrome and other abnormalities. It never occurred to us that anything could go wrong.

Drew: The radiographer looked at the screen and said, 'Hang on, something's not right.' She measured the length of the embryo. 'Oh, that's much too small, and there's no heartbeat. I'd better call the doctor.' She just walked out and left us. It was like being hit by a train. No one ever told us about the possibility of losing the baby at twelve weeks. We thought once we'd had a positive eight-week scan we were okay. We were ignorant at that stage, and it was a shock. I remember that day so clearly, walking back through the shopping centre to the car and feeling numb. Even today, walking through that mall brings back those nightmare memories. Whenever we had a bad experience at one of those ultrasound places, we never went back to the same one.

Caroline: We've worked through about one hundred of them!

I fell pregnant naturally four or five times, but none of them lasted. At the eight-week scan everything looked fine, and then at the twelve-week scan they were always dead. The second-to-last pregnancy was an ectopic, and I had to have a fallopian tube removed. When we found out during the ultrasound, we laughed nervously, 'Yeah, that'd be right. What else can you hit us with?'

The bottom line was that my eggs were too old. In the beginning we knew nothing about age-related infertility. When I was 38 a doctor showed me some startling statistics about my chances of conceiving and carrying a healthy baby to term. 'There's no question, if you're serious about having a child, you've got to start IVF now, although your chances are quite slim,' he said. I felt like I'd been shot. I was so shocked and upset. That was a wake-up call, and that's when we started.

Drew: For us, life prior to marriage had been a real dreamtime. This disappointment came as an unprecedented blow. We grew up in an era when women were told, 'Don't get married early. Go and get a career.' That was the message from the Women's Electoral Lobby and Women's Lib in the 1970s and 1980s, and that's great. I'm all for it, but there should be a clause explaining how it's much harder to get pregnant if you leave it too late.

Caroline: I would say to women in their twenties and thirties, 'Listen. Here are the facts about age and fertility. These are the statistics. Do with it what you will. But just know, if you leave it too late, you could miss out on having children altogether.'

Drew: When scientists learn to successfully freeze eggs they will change the world significantly. Every woman aged around twenty years will walk into the IVF clinic and freeze a load of eggs. I'd say to anyone, if you've got any spare bucks, invest in egg-freezing technology.

Caroline: I would definitely advise young girls to freeze their eggs because then there's not the huge pressure to find the right person early and have children. It takes the heat off.

Drew: It took a while for us to accept we'd have to do IVF. I thought, *That can't possibly be true. This isn't us. We'll be right.* I was surprised when we found ourselves in that position.

Caroline: I thought, *Everyone else seems to get pregnant naturally. Why can't I?* I've always had perhaps misguided confidence that I could do whatever I wanted. And my family has always encouraged me. I'd always been very lucky, and everything has gone my way until now. But getting pregnant has been my greatest challenge by a zillion times.

I didn't know anyone who'd done IVF, and didn't have anyone to talk to. At first I was embarrassed, and only told family and close friends. But, as time went on, I didn't care who knew. It was too hard to keep a secret, on top of everything else we were going through.

Drew: We were so naïve. We were absolutely sure we'd be pregnant on the first round of IVF.

Caroline: We thought that if they transferred a live embryo, of course it was going to work!

Drew: At the first egg harvest Caroline produced around 20, and lots of them fertilised. 'Fantastic, here we go. One go, and we'll be right,' we said. Little did we know!

Caroline: We did six rounds of IVF and none of them worked. One round was particularly devastating because we didn't even get to the transfer stage. We had six eggs and three fertilised, but they all died during the five-day waiting period. It seemed so unfair. I remember howling to mum and dad on the phone, 'I've been through all those injections, blood tests, and ultrasounds, and there's nothing to even transfer. How can this be?'

The drugs made me feel ghastly and I hated it, but I never missed a day of work. Some people succumb to the illness and depression, and take to their beds. I wasn't prepared to do that. I always kept going, because I didn't want to sit at home and think about it all day. When I was at work, I could be distracted — even though, some days, it was a battle to get there, because I felt like crap. I found the physical stuff easier to deal with than the psychological. My head was always my greatest challenge, the whole way through. I found it hard not to get really depressed. I'm usually a positive person, but some days I thought, *I can't go on. This is just so hard and it's never going to happen. What am I going to do? What's my life going to be about?* I'd always loved my work, but it just wasn't enough anymore. I thought, *I've got all this love to give. Who am I going to give it to? I have to give it to someone or something, or I'm going to die.*

In some ways, I wasn't strong enough to give up. If you keep trying, you delay having to face facts. It takes real strength to say, 'You know what? I can cope with not having a family.' I couldn't say that. I was going to get a baby, come hell or high water, no matter what I had to do.

I went to an IVF support group, and they may as well have been speaking Swahili. Several of them said, 'I can't do it any more.' And some had only done two or three rounds. I wanted to say, 'But you haven't even tried!' It just didn't make sense to me. It's not how I think.

I knew what it was like to have an amazing family and I wanted that, too. All our friends were having kids. When my closest friend told me she was pregnant, and it wasn't even planned, it really threw me.

By then we were at the end of our own frozen-egg supply and were facing a brick wall. One morning I woke up at five o'clock sobbing and saying, 'What am I going to do? You've got to help me, Drew. I've got to do something!' I was just beside myself. I was frantic and wretched. I kept saying, 'What am I going to do with all this love inside me?' That weekend we got Eric, our labradoodle pup, to ease the pain. We poured our love into him.

At the same time we started to consider egg donation. But the IVF doctor wasn't very hopeful. 'In this country, you obviously can't pay someone for their eggs, so it's virtually impossible to find the right egg donor unless you know someone or have a sibling who's willing to donate,' he said. I didn't know anyone, and I didn't have a relation who was suitable.

We considered advertising for a donor, but that didn't appeal to us. So we started down the international-adoption road.

We went to an information seminar about adopting from China, but it was bad timing. I was still very fragile, having just come out of hospital after an ectopic pregnancy and losing one fallopian tube. Dad had just died, and I was at a really low point. But we had to go to the seminar, because if you don't go you don't get another invitation. I was physically unwell and mentally wrecked.

Drew: I was not ready for the adoption idea. I hadn't given up on having our own children.

Caroline: Drew's viewpoint frustrated me, because the reality was I only had one tube and my eggs were crap. I needed a back-up option or else I would go nuts. I knew I could do adoption. I thought, *I love our dog Eric! I'm flipped over him and he's not even my species. Do you think I'm going to have trouble loving a child just because it's Chinese? Of course not!*

Drew: I really didn't like the whole adoption process. The Department of Community Services seemed very controlling, and there's so much red tape involved. When, after two or three years you get a baby, if you get one, the procedures don't stop there. DoCs is still in your life for the next twenty years, and that really didn't appeal to me. For example, you've got to keep the child's original name and commit to immersing yourself in Chinese culture and language, which they feel, from their experience, is the right thing to do. They're probably right. But you have to turn your world around to fit in with their scheme. It also costs around $20,000, which I guess isn't much compared to what we'd spent on IVF.

Caroline: I was keen on adoption because I couldn't stand any more unknowns. That's where Drew and I differed in opinion.

Drew: I didn't think we had any chance of finding an egg donor. If you look at all the ads in *Sydney's Child* magazine, the donor section is full of heartbreaking stories about childless couples pleading for donors. I couldn't see how we'd stand out from all the rest.

Caroline: We decided not to advertise in *Sydney's Child* because we thought we'd have Buckley's chance of attracting a donor. Ironically, it was because of those ads that our friend Neisha decided to donate to us. It was the weirdest thing. One day she happened to read some of those sad ads and thought, *Oh my God, this is something I could do to help*. She knows quite a few people in the same boat as us but, thankfully, she chose us!

I've known Neisha for about seven years, but initially not all that well, because she lived overseas for a lot of that time. When we first met, through a friend, we got on well. She happened to move up the road and we used to walk around Centennial Park together to get fit. You can really get to know someone thrashing around the park for an hour every morning. We are very fond of each other and have a similar outlook on many things.

When Neisha decided to donate to us she rang me at work one day and I nearly died of shock. It's not a phone call you expect. At first, when she said, 'I want to talk to you about something,' I thought she wanted to open a business with me, and I had absolutely no energy or interest in another bloody business. When she told me she wanted to donate her eggs to us, I came home and said, 'Drew, you're not going to believe this …' And he said, 'Quick, ring her back! Say yes!'

Neisha had finished her family and was prepared to have contact with any babies she helped conceive. This was an important issue to us. That's where having explored the adoption process was helpful, because there are similarities between adoption and donor, in terms of children wanting to know their genetic history. I read a lot of books about children of donors and adoption, and they all had questions like, 'Do you think I'm musical because my mum was musical? Do I have any predisposition to diabetes or heart disease?' It's our birthright to know our genetic history; most of us take it for granted because most of us know our parents.

Drew: My reaction to Neisha's offer was positive from the start. I thought it was a great idea. It was a gift from the gods. We were in a quandary at that stage. We really didn't have a clear direction, and were wondering what our life was going to be about. We were even considering doing aid work or being missionaries in Africa.

Caroline: Hardly missionaries. We're not religious!

Drew: Well, not missionaries exactly, but we wanted to do some good somewhere, to give our lives some meaning. By then, my eyes had been opened to what we could achieve as a family. While most people take for granted having a family, such a supposedly mundane aim of having kids and living in the suburbs became a holy grail for us. When we got Neisha's offer there was really no doubt in my mind it was a good thing. There were a couple of logistical things to figure out but, otherwise, it was fairly plain sailing.

Caroline: From the word go, Neisha was amazing. She said, 'You know best. You've done the research. You tell me what's the best way to do this.' She couldn't have been more incredible.

Neisha's husband Paul was great, too. He's a surgeon and pretty clear-cut about such things. He saw it as Neisha merely donating some cells, nothing more. He had it sorted in his mind in a minute and didn't have a problem at all, thank God. I know women who've wanted to donate, but their husbands have said, 'No. They're mine.' And their ego kicks in, whereas Paul said, 'It's a cell. It's a scientific process. Caroline's the one who'll carry the baby, and she and Drew will be getting up every night to feed it, not us.'

One of the reasons Neisha wanted to donate was because she's had so much joy from her own children and she knew people who would be good parents and were being denied that happiness. To her, it was never the big deal that it was to us. She says, 'I just hate it that you're so grateful, because I don't want it to be like that. It's no big deal.'

I still had to process giving up the dream of having my own biological children. Because I am so mad about my own family, I was sad their ancestry would not be continued. But I went to a brilliant seminar run by the IVF clinic, and heard an egg-donor couple and an egg-recipient couple tell their stories. The light switched on for me when the egg recipient spoke about her son and said, 'When Thomas was born, he wasn't mine and he wasn't hers. He was *him*. Thomas was Thomas.' When I heard that, suddenly all my concerns and fears became irrelevant. I think people can get very ego-y about it. You've got to let go of that, because it's not about us. It's about the children.

The IVF process is so medical. You become obsessed with your DNA and cells and genes, but it's not about that. But that's all you've got at that stage, so you become fixated on the technical aspect.

Neisha did just one cycle of IVF drugs. She was living in Balmain at that stage with her kids, but was supposed to move to Brisbane. Her husband Paul had already gone on ahead to start a new job, and she was going to finish the egg-donating process and then join him. But wouldn't you know, the month Neisha and I were supposed to start the drugs to get our cycles into sync, I fell pregnant naturally. When the doctor told me I burst into tears. I knew the pregnancy would fail like all the others — which it did — and it would only delay the process. Neisha was putting on hold moving her whole family to Brisbane because of me. It was disastrous timing. Fortunately, she was incredibly good about it.

Every morning we drove at about 5.30 or six o'clock across the city to her house. I looked after Neisha's kids while Drew gave her the injections or took her to the clinic. Then I went in with her on egg pick-up day. That was really difficult, because Neisha was overstimulated and in a lot of pain. She produced a phenomenal 22 eggs and felt puffy, bloated and sore.

I don't think Neisha was prepared for the discomfort of egg extraction. Unlike many clinics, Sydney IVF doesn't offer a general anaesthetic for the procedure, just sedation and a local. Even so, she found it wildly painful. Poor Neisha came back to our house after the egg pick-up and I nursed her while Drew was at work. She's not someone who complains, but you could tell she was in agony. I was upset and strung out, because she was suffering for me.

The good news was that, out of the 22 eggs, we got twelve embryos. We did three transfers with one each time … but none of them implanted. It was doubly hard this time, because I also had to ring Neisha each time and say, 'It didn't work.' I felt like I was letting her down. It was totally irrational, but I still hated making those calls. Neisha would say, 'Oh, I'm so sorry.' She even began to doubt the quality of her eggs. I also wondered if it was my fault. Everyone wanted it to work, and everyone was massively disappointed.

We also lost a couple of embryos when they were thawing out. Finally our doctor said, 'I think we should double the number of embryos we're putting in, because it's not looking too good.' So we transferred two embryos into me which didn't work, and then another two, and then another two. On the sixth round, I wanted to say, 'Just put the lot in, will you?' Because I thought there was no way it was going to work and I just wanted the agony over and done with. I reckon if I'd said, 'Put four in!' our doctor would have agreed, because it was clear he didn't have much hope either. We were getting the vibe from the doctors that we were in the too-hard basket.

Drew: The doctor didn't know why it wasn't working. There's still so little they really know. It's like Lotto. They can fertilise eggs in a petri dish and keep them alive for a few days and put them in the freezer, but they don't actually know why it sometimes works and sometimes doesn't. But, for some reason, I still had hope.

Caroline: Somehow 'Mr Hope' here was still hanging in. Meanwhile I'd received twelve of those phone calls that go like this, 'Hi Caroline, it's Sue from the clinic,' and I could tell by the tone of her voice exactly what she was going to say. By about the sixth time, I had it down pat.

'Hello, it's Sue from the clinic …'

'Okay, it's a no.'

'Oh, I'm sorry to tell you.'

'Okay, bye,' and I'd hang up and cry.

Then, out of the blue, I received a really upbeat call that went like this: 'Hi Caroline! How are you?'

As soon as she said, 'Hi,' I thought, *Hang on, something's different here*.

'Hi! How are you?' she said, 'Now your hormone levels are really high.'

'All right,' I said cautiously. 'What does that mean?' Of course, I assumed that something was wrong, because all we'd had to date was bad news.

'It means it's probably a multiple pregnancy,' she said.

My brain reeled. I didn't know how to process the news. We found out on Easter Monday and we were with some close friends. We told them, but immediately said, 'Don't say a word. Don't even smile.' We were so cautious about being happy. It wasn't until I was twenty weeks pregnant I actually believed there was any chance this would work. In hindsight, it would have been fun to be excited, but we were just so burnt by bad experiences. I felt sick with anxiety leading up to the twelve-week scan — because in the past, that's the time we'd always discovered the embryo was dead. But this time everything was fine. After that, I thought, *Okay, brain: you can stop worrying, and relax now*. I had trained my mind so well to turn itself off from adversity. I was six months pregnant before we began to relax a little and accept we were actually having babies. It was hard to believe after five years of disaster.

Drew: The chances of those two embryos taking, given all the others had failed, must have been one in 50 or something. If you're putting money on a horse at 50-to-one, you don't actually expect it to come in. We'd had twelve failed rounds of IVF. Finally we'd hit the jackpot, but still we went through periods of panic and nervousness right up until the birth.

Caroline: I was booked to have a caesarean, but it turned into an emergency procedure when my liver began to pack up. It was overworked and underpaid! There was also meconium in the amniotic fluid, which meant the babies were distressed. We didn't have time to muck around. We had to get the twins out quick smart. They pressed the emergency button, and it was all on. They were going to do more tests and then the doctor said, 'No, we're not taking that risk. We're getting them out now.' I'd already had a steroid shot the week before when we'd had a false alarm. So we knew their lungs would be strong enough. Our baby girls, Jessie and Eve, were born six weeks premature.

Drew: I couldn't be in the operating theatre while Caroline was being prepared. I could only watch through a little window. But I was there when they pulled the babies out.

Caroline: Finally they let Drew come in, but then it all went wrong, because the spinal block didn't work. So he had to go out again while they gave me the general anaesthetic, and then he came back in.

Drew: A lot of people say that the moment you lay eyes on your babies you think they're the most beautiful things you've ever seen. Well, not in our situation, because this was in a full-on operating theatre. There must have been fifteen people working in there, and more machinery than you can imagine. And in the middle of it all, my wife was being chopped up on a table. I saw these little baby things that looked liked skinned rabbits being pulled out, but I was more distracted by seeing my wife out cold and cut open. Then the babies were whisked off for tests. I wasn't allowed near them, because the doctors and nurses were trying to ventilate them. It wasn't the most relaxing situation. It's very different from the manger, straw, and the three wise men scenario.

Caroline: Well, there were wise men …

Drew: Sure, there were plenty of wise men, but they weren't bearing gifts — they were bearing invoices!

By the time we were reunited with the babies, it was almost an hour later, and they were in humidicribs and rigged up with drips and splints. And we still couldn't hold them. The poor things had stents stuck in their tiny hands, which were no bigger than my thumb. It was unbelievable.

Caroline: The day Neisha came to the hospital was very emotional. I was excited for her to see Jessie and Eve, but there was also the question in both our minds of how she was going to feel about them. Neisha says she feels the same way about them as her godchildren or nephews and nieces. She feels a connection with these children, but not that they're hers. She didn't think, *Oh my God, I want to take them home.*

Not that I ever thought she would. From the word go she was very pragmatic and clear. If there'd been any doubt at the start, I wouldn't have gone ahead. Otherwise it could have been an incredibly fraught situation. Neisha says she could never be a surrogate mother because she feels the bond comes from carrying the babies, giving birth, and breastfeeding.

The hospital staff said it would be about three weeks before we could take Jessie and Eve home. We had to wait until they could moderate their own temperature, be bottle- or breast-fed rather than with a feed tube, and were consistently putting on weight. They were so tiny. Even when we took them home they were only about 1.6 kilos each.

Once we had them home it became overwhelming, because we were up all night. I found it very hard, and was anxious all the time. I didn't know what I was doing. Then they got reflux and were screaming in pain. I had no idea about anything until I got some advice on getting them into a sleeping and feeding routine.

I breastfed for about six weeks, then we went on to bottles. We were doing sixteen bottles a day at different times. I was probably over-feeding them, and often they were sick all over me. It was frantic. My sister and I were going like bats out of hell 24 hours a day, with mum doing the bottles and the washing. Whenever Eve and Jessie were awake, they were screaming in pain. It was hell in those first few months. I was totally inexperienced, and so tired and stressed. My children were screaming in pain and I didn't know how to help them. They were also overtired, because I didn't know how to help them sleep. It spiralled out of control into chaos.

Drew: I remember saying, 'Crikey. We've gone through hell for six years to get here, and now this is hell as well.'

Caroline: Sometimes I thought, *I'm hating this*, and I felt guilty about that, especially as an IVF mother. I was supposed to be joyous, but I was so tired, I couldn't even speak and just cried all the time.

It never occurred to me just how physically draining it would be to have two babies — constantly picking them up, feeding them, changing them and rocking them. My body was in pain.

I'd finish feeding one and put her down and then the other would start, 'Waaaah,' and I thought, *Please God, no. Please God, no*.

Neisha lives in Brisbane and, to be honest, I would never call on her to help, she's done more than enough. Anyway what would I say? 'Could you come and be tortured?' Fortunately, I had my mother and sister staying here, who went straight into action mode. They're family, so you don't have to talk. You can walk around in your undies and nobody minds. They were spectacular. My sister split the night shift with me. Who in their right mind would do that? They looked after me, too. Mum got up at six o'clock and did the morning feed and made me breakfast. I was terrified when she went home to Melbourne.

At two o'clock in the morning, as I was pacing up and down with Jessie, who was screaming from reflux and I knew I had another four hours of it to go, I used to think, *I'm so lucky to feel this crappy, and at least it's not triplets!*

Fortunately, it all turned the corner when we found out how to get them to sleep and got them on medication for reflux. Now they're bigger everything is coming right.

It's been one hell of a road! But in a few years time, when it evens out and they're playing happily together, it will all seem like a distant memory.

We'll tell the girls about their conception as soon as they're old enough to understand. They will probably call Neisha 'Aunty', and her children will be like their cousins. That works for all of us. There will be a special connection with them, which is nice, because we have very small families.

Our girls will always know Neisha is their genetic mother. Her own kids already know, although they're too young to really understand. They know their mummy gave me some eggs. It's all very open. There's nothing weird about it. That's just how it worked. Some people donate organs to keep someone alive. Well, that's nice. But two new lives? Neisha's given us a family — it's so big, it's almost inconceivable.

Drew: I don't get hung up on the genetics. I feel we're the custodians of two little people. Our job is to put them on the best possible path, and then one day they'll go off and do their own thing.

Caroline: We're theirs, but they're not ours. Even if they were genetically mine, I still wouldn't own them. They own me. We're here to make it as good for them as we can. That's how I sincerely feel. Jessie and Eve are so strongly their own identities; it's not even a question of whom they belong to.

Drew: It's great, because it's all ended well. Our story was so close to going completely differently.

Caroline: We came close to total desolation, but we hit the jackpot, having two. Now that we've got the night nurse and I'm sleeping and back to being myself, every moment is pure joy. Every time I look at Eve and Jessie, I think, *Oh, they're so divine. They're better than perfect. They're miraculous.*

CHAPTER 4

Too Little, Too Late
–Di's story

In her late thirties, Di started asking herself, 'What am I going to do with the rest of my life?' She really liked what she did, but it had become less of a career and more of a job. She and her husband, Don, left the city for the country almost ten years ago because they thought it would be a great place to raise a family. Now they're looking at plan B.

Don and I have been married for eighteen years. We met in Orange in New South Wales, where I did my journalism cadetship. We went back to Sydney for a while, and delayed having children because we wanted to travel, build a house, and give our careers a go. I wasn't keen to have a family in Sydney, so when we were finally ready we moved to Albury, on the Victorian border. By that time I was 33 and Don was forty.

I have hypertension, so I can't take the pill, and I used a diaphragm for years. I look back now and realise I was really naïve, because we thought the contraception was working really well!

We gave up the diaphragm and I tried naturally for about five years to fall pregnant. I was disappointed it wasn't working, but still positive, and thought, *Well, it will happen eventually.* I remember being with my sisters in a bookshop in Sydney and picking up a book about natural conception. They both looked at me strangely and said, 'What are you doing?' I'd never discussed wanting a family with them until then. 'Well, we're trying to have a baby,' I said. I listened to myself say that and realised, *I am actually quite serious about this.* I was about 35 then.

Don didn't seem too concerned either — although a couple of times he said, 'We should do something about this.' I guess he was waiting for me to take the lead. In hindsight, I was in denial that anything was wrong.

For some reason, I found it difficult to raise the issue with my GP. I'd never had any menstrual pain or ovarian cysts, and there was no family history of reproductive problems. But finally I said something, and she referred me to a specialist for a laparoscopy, hysteroscopy, and all the other usual tests. There didn't seem to be any detectable problems. They also checked my husband's sperm and that seemed fine, too.

As a first step, the doctor put me on Clomid to stimulate my ovaries, which made me feel like I had a belly full of marbles. We did a few artificial inseminations (IUI) with my husband's sperm. IUI is not too invasive: it's a bit like a pap smear, slightly uncomfortable, but not a major hassle. I thought we were really fortunate because we were living in a country centre which had its own clinic, and we didn't have to travel to Melbourne. The clinic was literally around the corner from where we live, which made it very easy, and I felt pretty confident we would succeed. However, after six months of IUI and no pregnancy, we knew IVF was our next step.

Even then, I was still in disbelief that there was something seriously wrong. I thought, *Whatever it is, they will be able to tweak it and it will be okay; it will be pretty straightforward.*

I had a good response to the IVF drugs — in fact, a little too good. I produced 23 eggs, fourteen of which fertilised. We had to wait until my symptoms from the hyper-stimulation subsided before transferring a blastocyst into my uterus. I would get up every morning and look in the mirror to see if my tummy had deflated. I'd walk into the surgery, and the nurse would look at me and say, 'No, you're not ready yet.' I really wanted to do a fresh transfer rather than a frozen one, and I was willing myself to get better fast. Fortunately, I recovered quickly, and they put one blastocyst in me on day five and froze the others. During the transfer the clinic played weird elevator music and dimmed the lights in the hope of creating a calm atmosphere.

I didn't have any huge expectation, because they always warn you the first time is the test cycle to see how you respond to the drugs. But it was hard not to get too excited.

Ten days later, I was at work at the newspaper when the nurse from the clinic rang. 'You're pregnant!' she said. 'You're kidding!' I screamed so loudly the editorial secretary, who knew who I was speaking to, beamed and said, 'Quick, ring Don!'

It was really hard not to tell everyone, and there were so many expectations, particularly once they knew we were doing IVF. Everyone was willing us to succeed.

We went in for a seven-week ultrasound, and it was a really weird sort of afternoon. The baby's heartbeat was very low, only 85 beats per minute when it should have been between 120 and 140 bpm. I didn't know that at the time, but there was something in the doctor's response that rang alarm bells. He took a really long time looking at the scan; and Don, who was sitting next to me, grabbed my hand, because I think he felt it, too. The doctor mentioned something about the heart rate, but didn't express any major concerns. He said, 'I'd like you to come back for another scan next week.' I walked out of there and must have had a strange look on my face, because the receptionist said, 'Are you all right?' and I said, 'I think so,' but I wasn't sure. Because the doctor hadn't said anything definite to alarm us, I spent the next week as per usual and even started to get morning sickness and feel pregnant. We went in the following Wednesday for another ultrasound, and within a minute the doctor said, 'I'm sorry, the baby's died.'

It was like a blur after that and I didn't hear anything else. Don told me he went on to say, 'It's okay, we'll go again, and you've had a good response,' and all the things you expect people to say.

Don was broken-hearted, too. We came home and cried. For some unfathomable reason, I went back to work at midday, as if on autopilot. The editorial secretary said, 'Are you all right?'

'We've lost the baby,' I said.

'What are you doing here then?'

'I just didn't know quite what else to do,' I said. I was stunned, and it all seemed very unreal.

We expected I would miscarry naturally because it was so early, but within a week I was in a lot of pain and had to go in for a D&C. My doctor was away but, fortunately, a very kind and gentle obstetrician did the procedure for me. After that I didn't get my period for five months, which made me increasingly frustrated and depressed, because we couldn't start another cycle until I started menstruating again. It was probably at that point I started to think, *Well, maybe this might not be quite as simple as I thought.*

Most of our friends were supportive, but didn't really know what to say. My in-laws, who are in medicine, were quite clinical and said, 'Oh, well, you can just go back and try again,' and rattled off statistics about miscarriage rates, which I didn't find very helpful. Meanwhile a colleague's wife fell pregnant on her first IVF cycle, and they've since had the baby. I started to get a bit gloomy, and it probably coloured the way I approached the rest of the treatment.

I finally got my period again in February, and we went back and did two frozen transfers in April and May. Neither worked, so we did another whole stimulated cycle and egg pick-up in August, but we ended up with only two embryos to transfer and none to freeze.

I was quite proud that I was able to inject myself and became quite proficient at it, but poor Don would have to leave the room. There were a few humorous moments when I panicked as to whether I had given myself the right amount of medication. It was much better when we switched from using ampoules to the dial-up pens, which were heaven in comparison. Once I accidentally stabbed myself in the finger with one of the draw-up needles, which looked like a big darning needle. I had a huge bruise on my finger for weeks, and people kept asking, 'What have you done to yourself?' I said, 'Well, you wouldn't believe me if I told you.' It was bad enough having bruises on my stomach, let alone where everyone could see!

Don used to laugh when he'd see me every morning with my little medical pack, standing in the dining room stabbing myself in the tummy before going off to work. In a strange way, I found all that mechanical stuff comforting, because I felt I was moving towards my goal.

What I found difficult was doing nothing during the two-week wait between the transfer and the pregnancy test. It would send me around the twist when I started to dwell on it. Whereas when I was distracted by injections, scans, or the egg pick-up I was occupied, which made it easier. It's when all of that stopped and it was in someone else's hands that I got nervous.

By this stage we were thinking about trying another clinic. We were concerned that the embryos being frozen weren't the best quality and wouldn't necessarily be frozen elsewhere. I suppose part of it is because the clinic here is not in a major city with a huge research program to back them up and make their techniques cutting edge.

We were surprised to find that different clinics have different policies on when to transfer and when to freeze. For example, Melbourne IVF does transfers and freezes after two days, unless there's a request otherwise, whereas some clinics transfer the blastocyst after day five. You quickly realise all clinics are not the same, and it's hard to know which offers the best procedure. We decided to have some of our treatment down in Melbourne, while doing my scans in Albury, so I didn't have to travel too often.

After the failed cycle in August we did another in December, and ended up with ten embryos to freeze, although we lost two of those in the thawing-out process. We transferred the remaining eight back into me during four cycles — putting in two each time.

In some ways I wasn't surprised when I got the phone calls saying, 'Sorry you're not pregnant.' After that first miscarriage, I'd become attuned to knowing when it hadn't worked.

In a number of cases, I've had my period before the blood-test results have arrived, which is what happened the last time. I didn't even bother having the blood test this time around. I rang the nurses and they said, 'Are you sure you're not going to do a blood test?' And I said, 'Yes, I can guarantee I'm not pregnant. I did a home pregnancy-test three or four days ago and that was negative, and now I've got my period.'

I appreciate they're trying to cover themselves in case the pregnancy is an ectopic, but you feel like a fool going in for a blood test when you already know the result.

Don would always be very concerned about me because I would get so upset after each failed cycle. I would do a lot of fairly serious crying. Crying really seems to scare some blokes. Whereas for me, crying was a welcome release and something I needed to do.

We've now done thirteen cycles: four stimulated cycles and the rest with frozen-embryo transfers. After we ran out of frozen embryos, Don and I decided not to continue. We came to the decision at the same time. One day, Don said, 'I just don't want to see you going through this anymore.'

This is the third time I've said I'll give up — only this time I mean it. I've had enough; I don't want to do this anymore. I've known this day was coming and I've hoped like hell it wouldn't come to this. When I did the last stimulated cycle I'd virtually made up my mind this would be the last, because each stimulated cycle has been increasingly more difficult. While I haven't had a really rough time like some people, the last one was very uncomfortable and I kept wondering, *What am I actually doing to my body?*

At the moment I feel quite balanced hormonally, and I haven't felt like that for a long time. It's taken six or seven months for my body to get back to normal. I think you underestimate how long it takes for your system to settle down.

We didn't consider donor sperm, because there was no indication Don had a problem. And as far as donor eggs go, they are really hard to find. There wasn't anyone we knew. I've got younger sisters, but they haven't had their own children yet. Both of them offered to have the baby for me, but it would be different if they had their own families. It was just way too much to ask. My youngest sister is getting married next year; this will be the first wedding in the family since ours eighteen years ago. She'll be 36 when she gets married, and I hope like hell they get pregnant straight away. Nothing would give me greater satisfaction.

I don't think there's anything quite the same as the strain of IVF on a relationship. Fortunately it has made us stronger, but I can see how it could easily go the other way, too. The important thing is to keep talking, but not when you're angry or upset. You have to try to keep it on an even keel, and that's not always easy to do. We've had blues and heated discussions about whether or not to continue IVF. We've walked away from it for a couple of days and then come back to it again. It's like constantly planning a battle, and you have to keep reassessing things with what little information you have and asking, 'Is this the best path?'

Doing IVF can sometimes take the shine off lovemaking, too. But in some ways it makes you work harder at those things, and you become more thoughtful about creating opportunities away from all the IVF crap.

Last night, I was talking to some of the other women on the internet support chat-room about how it's so easy to get swallowed up by the whole IVF machine. Each time you tell yourself you won't do that, but then you get involved again and it becomes everything. IVF becomes the thing you eat, sleep, and think about all the time, and you have to make a concerted effort to retain a normal life. I get cranky at myself because I've lost quality relationships with friends and colleagues during the past few years, because I've been so consumed with the process. I've probably burdened Don with a lot more than I should have. That's why it's imperative to find a good counsellor so you don't lump your partner with the entire load.

Sometimes you have to speak to a few counsellors before you find the one you click with. I used to talk on the phone to a particularly good one down in Melbourne. There's also an outstanding grief and loss counsellor here in Albury. She's been excellent at this latter stage, because now we're mourning the loss on a number of fronts — not only the baby we lost, but also the opportunity to have our own child.

Handing over power and control has been really hard for me, particularly because they haven't been able to explain why it hasn't worked or how we can solve the problem. I spot about a week before my period's due, which may have something to do with it. There's got to be an implantation problem, or it's something to do with the quality of our eggs and sperm. The trouble is, after you go through years of trying and you get to my age, you don't know if you're dealing with an inherent problem or if it's age-related or both. Fortunately, the counsellors have helped me to stop asking those questions, because it has at times sent me round the twist. In some ways it's probably a good thing we don't know, because it means neither of us is to blame. It's been part of my healing to realise I can't always control things and it's okay not to have the answers, and not everything can be fixed.

In the beginning I thought it would work, because the doctor said, 'There's nothing to indicate why it shouldn't work.' It was also the vision of a baby and a family that kept us going. My parents don't have grandchildren — which is tough on them, too. I would love to see them as grandparents.

Dad is quite a sensitive soul and he recently said, 'You know, Di, you haven't spoken much about the last few cycles,' and I said, 'Mainly because it's all just so damn sad. I'd be leaping out of my skin and round the corner telling you if there was something positive to come out of it, but there hasn't been, and there hasn't been for some time.'

Parents always want to make things better for their kids, and I've had to help them realise there is nothing they can do other than just be there. They have been very sweet and even took me down to Melbourne for one transfer when Don was away on business.

I'm getting older, too, which has been banging at the back of my mind for two years now. I'm forty in a couple of weeks. On the weekend I saw a girlfriend who has kids who are ten or eleven and she said, 'You're not having a party?' and I said, 'No, I'm not.' I've been quite firm about that. Not having a child has really spoilt it for me in a lot of ways. I don't feel like celebrating. I would rather go away somewhere quiet like the Margaret River in Western Australia, just the two of us.

Don and I came to the idea of adopting at slightly different times. I told him I wanted to adopt and he said, 'I don't know whether I could do that.' It took him longer to get his head around having a child that wasn't biologically ours. But now he understands my hunger for a baby and he wants it, too. I wanted to have a baby of my own, and my blackest time was having that desire when it was becoming increasingly obvious IVF wasn't working. I still have a strong urge to be a parent, and that's why adoption is the right answer for us.

Adoption gives me some sort of hope to hold on to. We're just in the beginning stages. It could take up to three years until we get a child. Adopting from China is our only chance because Don is 47 and some countries have an age limit. There's already a Chinese adoptee in our family: my husband's younger sister, who lives in Scotland, adopted little Ailidh a year ago.

I am quite excited about adopting and a bit terrified, too. It's a similar feeling to when we started IVF. In some ways, adoption helps us take back control. We can now say, 'Okay, I accept I can't do anything about the past. We've tried and we've thrown everything at it. I know what the boundaries are here, I know what I've got to work with, so let's just go for it.' Adoption comes with its own set of problems, too, but if it's a choice between not being a parent at all and adopting, there's no question.

It's incredibly cruel when IVF doesn't work, but there are people who have been through far worse situations. It's something you wouldn't wish on your worst enemy. I get cranky when I read the *Herald Sun* — I call it the miracle-baby newspaper. It always prints the success stories and not the other side — it's irresponsible. Obviously, we left it too late. Somebody should have kicked us in the pants somewhere along the line and said, 'You need to do this now.'

When IVF works, it's absolutely fantastic. One of my friends had a miscarriage and then had IVF twin girls who are just delightful. Other friends, who are much younger than me, got pregnant on their first go at IVF and their little bloke is lovely, too. So you don't want to scare the pants off people; but at the same time what irks me is the perception out there that, 'We can delay having children until the last minute and it will all be fixed by IVF.' I want to shake people and say, 'Look, it might work for you, but it just might not, either. Do you really want to go through all that and still end up childless?' Not everyone has the option to adopt, either; we're just lucky we have the funds to pay for it.

I'd like to think that, in the future, IVF will become less about trying to extend people's fertility into their forties, fifties or sixties, and more about trying to find the answers to people's conception problems when they're younger. I'd like people to think about how important being a parent is, and factoring that into their lives earlier. It's so easy to get caught up in, 'We've got to get a mortgage, we've got to do this and experience that before we settle down and have the family.' Without sounding too trite, being a parent is really the most important thing you can do. And for anybody doing IVF, I'd say, 'Look after yourself, don't let it run you. You've got to look after yourself, otherwise you'll fall in a heap.'

Postscript

More than a year has passed since I first told my story, and a fair bit has happened. My sister has married, and fell pregnant on their honeymoon. The baby is due in two months; it will be my parents' first grandchild. Don and I have just completed our interviews for adoption. Unfortunately, the waiting period has lengthened considerably due to increased demand. I did think about IVF again, with a donor egg, but only for a short time. We've moved on and are content to wait for our adopted daughter, who we hope to meet in late 2009 or early 2010.

CHAPTER 5

Towards the Light
–Erika's story

When three ectopic pregnancies rendered her fallopian tubes useless, Erika was devastated. She had dreamed of being a mother since she was a child. Erika's only option was IVF, but her path to motherhood was pitted with grief following the ending of two important relationships, the loss of her younger brother, and her own near-death experience. But she never lost sight of her dream. IVF helped her realise her goal, along with a lot of soul-searching and self-nurturing.

I was born to be a mother. When I was six years old, I used to strut to the shops in my yellow flares with my baby doll in a pram. I'd stop at the lights, waiting to cross the road, fussing over my doll. I thought everyone in their cars was watching me and believed I had a real baby.

My aunty had twins when I was eleven, and I used to go to her house every day during the school holidays and pretend one of them was mine. I didn't know much about babies, but loved everything about them. I wanted to be a mum from the time I was three, when my brother was born.

At 24 I was in a long-term relationship and fell pregnant. But my partner decided he wasn't ready and my family also told me

I wasn't ready to be a mother. So I had a termination. When I came out of the anaesthetic, they said, 'We didn't get anything. There was nothing there. You've got to have an ultrasound.' The ultrasound showed an ectopic pregnancy. There was an embryo in one of my fallopian tubes and I had to go to hospital.

I had no idea what was going on or what the implications were. The doctors were amazed I had a nine-and-a-half-week-old embryo in my tube and it hadn't burst. I didn't feel any pain. I just thought the twinges were all part of pregnancy. In fact, I felt fantastic and, although I had morning sickness, I felt proud and grown-up. I don't think I comprehended that this was an abnormal pregnancy.

Fortunately, they didn't have to remove my fallopian tube. I saw an obstetrician at the old Royal Hospital for Women in Paddington, Sydney, who was doing a new type of laparoscopy. He lasered the tube open, removed the embryo, and closed it up again. He'd done about twenty successful procedures so I was naïvely confident. It was keyhole surgery through the navel with a couple of other little holes and a few stitches. I woke up after the operation some time in the morning and everything seemed fine.

But a part of me felt numb and sad to have lost a little soul. Another part of me felt bamboozled by the whole situation. I'd left home the morning before to have a termination, thinking I'd be home by lunchtime, but instead I was in hospital the next day recovering from major surgery. I'd never heard of an ectopic pregnancy, let alone had an operation.

Later that night, something went wrong. My stomach was contorted with cramps and I was deathly pale and sweaty. After checking me, the doctors rushed me in for emergency surgery in the middle of the night. I haemorrhaged and lost almost half my blood, and had to be given a transfusion.

I flat-lined and I remember seeing a white light. I felt like I was dying. Going towards the light, I felt euphoric, weightless, and full of love. I thought, *This is incredible; this is how life should be*. But I questioned whether I was ready to go yet. I felt I had the power to choose life or death. The first person to come to my mind was my brother. I saw his face clearly. Then I thought of mum, dad, and my partner, and thought, *Maybe I'm not quite ready*. So I came back.

That experience made me respect life a lot more. It made me aware of the power of love and our choices. I was excited to have seen the light, but also shattered that one of my fallopian tubes had been removed during the operation. Now, with only one tube, my chances of becoming a mother easily had decreased.

I was drugged up with pethidine shots every four hours to numb the physical and emotional pain. I felt frail and lost heaps of weight. My body felt like a fragile shell in need of protection. At that stage I didn't know about meditation or alternative healing therapies, so I stuck with my doctor and conventional medicine.

My second pregnancy was six years later when I was thirty. I was seeing a younger guy, and having a child with him wasn't a viable option. I had an ultrasound because my doctor said, 'Considering your past history, let's check this out straight away.' It turned out to be another ectopic. Once again, I had the same type of laparoscopy but this time with a different obstetrician. Obviously, I was concerned I might lose the second tube, but my obstetrician assured me the technology had improved. He gave me a 99-per-cent guarantee I would have a usable fallopian tube at the end of the procedure.

But something very tricky happened with that one. My body didn't recognise that the embryo had been removed. A few cells remained attached somewhere, and my body thought it was still pregnant. I had to go in to hospital every second day for a fortnight to have chemotherapy to kill off those cells. It was pretty intense. I had a series of blood tests, and each time the lab technicians would come back and say, 'Sorry, it's still registering as something there. We need to give you another jab.'

This time I was really annoyed at myself for getting into this situation. It made me sit up and look at what I was doing. Why was I in this relationship? Was I really committed to becoming a mother? I had bright, fire-engine-red hair, wore wacky, vintage clothes, and partied hard. I'd just been to San Francisco and discovered a different side to myself, and was having fun exploring that. In hindsight, I was probably more ready to become a mother at 24 than at thirty.

Fortunately, I didn't lose the second tube, so I didn't give myself too hard a time. I just got on with it. I was working for the Australian Film Industry Awards, which was really exciting, and I buried myself in my work and moved on.

A couple of years later I was single again and in a new job. My company sent me to London as the marketing manager for the Mind, Body, Spirit Festival. I was suddenly in the world of healing and alternative therapies, and meeting lots of gurus and healers. That was a good place for me to be, and that's where I met the man who asked me to marry him. He was a yoga teacher and sold musical instruments, arts, and crafts from Asia. He supported orphanages and was an all-round decent guy. We decided to make a go of it.

I came back to Sydney for ten weeks to sort out my affairs. While I was here, I thought, *I might as well get myself checked out*. I didn't like wondering whether or not I could get pregnant. I had the iodine test in my remaining tube and was given the all-clear.

After that, my partner and I met up in Nepal. We were working at an orphanage when I got a phone call telling me that my brother had killed himself. I was gutted. We came back to Sydney and dealt with the pain, sadness, mess, and confusion. I also felt guilty because I hadn't been there for my brother during his darkest hours. I'll always wonder whether I could have helped him if I'd been in Australia.

Six weeks later, we found ourselves working at the

Glastonbury Festival in England. We were stuck out in the middle of nowhere in a chaotic sea of tents and people and musicians. I thought I could lose myself in the mayhem and try to bury my grief. Then, one day, I woke up and I couldn't walk. I was in so much pain. I felt constipated, bloated, really sick, and toxic. There was a constant beat of music from the festival, and I couldn't find any peace. For two days I lay in the tent with a stream of people coming in to do reiki on me. As soon as I got back to London I went straight to a doctor. I was still grieving for my brother and felt very disconnected from reality.

The doctor sent me off for a scan and, sure enough, it was another ectopic pregnancy. I was rushed by ambulance to hospital before it burst. I was in so much pain I could barely get off the bed. I was on all fours, screaming in agony. That was the only position I could be in. It was really horrible and very frightening, to be so far away from home, with no family around me. My partner adored me, but I wanted the support of people I'd known all my life.

As I clenched the bar at the end of the hospital bed and cried with pain, I thought, *Why is God doing this to me? My brother's just committed suicide. Do I have to deal with this as well?* I was confused and angry and wondered, *Why me?*

I had emergency surgery and they reefed my fallopian tube out. The doctor said there was no point trying to salvage it. 'There's no way you're keeping this one. You've already had one pregnancy in there.' When I woke up I was devastated — absolutely devastated.

I found English hospitals and nurses really scary. It was like something out of an old movie. There were long corridors of iron beds and severe-looking nurses with flapping, 'flying nun' hats. I was far from home, and the hospital was miles from where we lived in London, making it hard for anyone to visit. That's when I finally fell to pieces. I hadn't had a chance to tell my parents because it all happened so quickly. I thought, *My God, I've just been through this huge ordeal, and my family don't know anything about it.* That was the turning point, when I decided to come back to Australia.

My ordeal in hospital and my brother's suicide suddenly put my priorities into razor-sharp perspective. Now I wanted a baby more than anything, and nothing was going to stop me. I was 33 and the pressure was on. I said to my partner, 'Right. We need to do IVF.' I swung into action, got on the case, and did the research because I wanted to get it sorted before I was thirty-five. Who knew what the odds were going to be? I knew I had nothing wrong with my fertility, but I didn't know how I'd cope with IVF.

I thought, *If we've got to do IVF, I want to do it in Australia.* Losing that fallopian tube had left me with a lot of scar tissue and pain, and I didn't like my experience with the English medical system. Now the pressure was on my partner to come back to Australia with me. But he had his business, his two dogs, and his yoga, and wasn't ready to move to Sydney to have a child he wasn't sure he was ready for.

So I stayed in the UK and prayed he'd change his mind. For eight months we tried to get on with our lives, and then I just crumbled. I realised he wasn't the one for me, and England wasn't the country for me. I had so much grieving and so much healing to do around the loss of my brother and my tubes. I wanted to be back home, surrounded by family and old friends.

I arrived home single, but still determined, one way or another, to have a baby. I researched IVF by reading books and talking to doctors. Since my first ectopic pregnancy I'd always known IVF was an option. Anything that helps people gain some sort of normality in their lives or helps them realise a dream is amazing, in my opinion.

I understood where I had to be and what I needed to do, and got right onto my pre-conception care. That meant addressing my diet, taking supplements, giving up cigarettes and party drugs, and cutting back on drinking. It was the turn of the century, so it felt very cleansing; it was like a re-birthing for me. I was looking after a friend's house at Bondi, so I had time by the beach to re-connect with myself.

And then I met Jonathon. We'd known each other for a long time, but hadn't seen each other for eight years. For the first time we were back in the same city, at the same time, without partners. All of a sudden, we realised there was a lot more chemistry between us than we'd previously acknowledged. I told him about my brother's death and the drama with my fallopian tubes. But I didn't tell him about my burning desire to have a baby — I didn't want to scare him off! After being together a few months, we were lying on the beach and he said, 'I don't want kids. I don't really like kids,' and I thought, *Well, you're definitely not the one for me then*.

But the more we got to know each other, the more intense it became. We fell in love and stayed together. About two years into the relationship, he looked at me one night over dinner and said, 'Let's have a baby. Let's go for it. Let's sign up for IVF.' I said, 'Oh my God! Okay!' I was primed and ready to go.

We chose a clinic in the western-Sydney suburb of Hurstville. It's run entirely by women, and I figured they'd have more of an emotional connection with their female patients. It was quite a long drive, but it was beautiful because every visit became a purposeful journey. It wasn't something I just fitted into my lunch-break in the city. It was a real excursion, and gave me time to reflect and meditate on what we were doing.

Jonathan was excited and ready, too. The whole IVF process didn't faze him. He said, 'It's going to be really simple and uncomplicated. We both know we are fertile, so let's just get pregnant. Let's have the chubba.'

We went to the clinic, and he went off into one of the little rooms and did his business and came out with his offering. It's bizarre for the men, because while we women sit out in the waiting room we know exactly what they're up to in their little cubicles. It's all so clinical, in a way.

First, I took the nasal spray, because I'm not a big needle fan. But the spray irritated my nose and I was worried I'd sneeze everything out. I thought, *There is no way I can do this three times a day for six weeks.* So I went back in the next day and got the injections. Initially, Jonathan did them, but I realised he wasn't going to be around for all of them, so I had to get a handle on it quickly.

Before injecting, I used to think about the baby to give me courage. It became a beautiful daily ritual. I visualised being pregnant and having a baby in my arms, which I'd dreamed of since I was three. Those dreams were closer to coming true, and I didn't fear an ectopic pregnancy anymore.

The day before the egg harvest, I found it really painful walking to yoga. Both my ovaries were so full of eggs. I felt puffy and sick from the treatment. I was working in a massage clinic in the city and had lots of lymphatic drainage to ease the bloating and puffiness. At the same time, I felt full of the ability to have life inside me, and I tried to embrace the process as positively as possible.

The next day I went to the IVF clinic and had the egg extraction under general anaesthetic. When I woke up they said, 'Look at the number we've written on your right hand and it'll tell you how many eggs we got.' I looked at my hand and it had 22 written on it. Twenty-two eggs! No wonder I couldn't walk the day before. They said, 'Well done! That's fantastic!'

Sixteen of the 22 eggs fertilised, which meant we had plenty to play with. I felt proud and excited. Then we had to decide whether to have one, two, or three transferred into me as soon as they'd fertilised, or choose to wait five days until they became multi-celled blastocysts. They said the success rate was greater with a fertilised egg if it had survived five days in the petri dish. So we opted to wait and have one blastocyst transferred.

I felt battered and bruised from having my ovaries pierced 22 times with large needles. My insides felt swollen and tender, even violated. I looked pregnant, but I was just puffed up from the drugs and procedures. I tried to carry on my regular life, but it was difficult.

Those two weeks waiting for the results of the pregnancy test were among the worst of my life. I was massaging in the city on the day the results were due. Jonathon rang every half hour, asking, 'Have you heard? Have you heard?'

'No, not yet.'

'Ring them! Ring them!'

'No!'

Eventually I rang. 'No Erika, the results aren't back.' So I rang Jonathon, 'The results aren't back. Please don't hassle me. I've got a massage to do.' There was so much pressure waiting for that phone call. It was nerve-racking. It was like waiting to see if I'd won Lotto.

Finally, the clinic called to say it hadn't worked and I wasn't pregnant. I rang Jonathon and then had to go straight in and massage a regular client. I was crying and trying not to make any noise while the tears streamed down my cheeks. I kept moving my head so the tears wouldn't drop on my client's back. That hour was the longest of my life. I couldn't get out of there quickly enough. I ran into the yoga room, crouched down, and bawled my eyes out. I was absolutely devastated. Unfortunately, the client walked in, and I thought, *Oh no, he's going to think I'm crying because of something to do with him.*

There and then, I decided I was not ready for another round of IVF. I needed a break. Taking the fertility drugs, and making and harvesting all those eggs had taken its toll. The added pressure of being pregnant straight on top of that would have been too much. I needed to get back to myself. I decided to store the other embryos, which I now think was the best move. I thought, *Why would a little embryo want to implant itself in here, when it's raw and tender from all those drugs and hormones? I don't blame it for not wanting to make this its home.* Finally I came to a place of acceptance and letting go.

For the next six months I worked on myself again and tried to make my uterus a beautiful, welcoming, and warm place where a little embryo would want to live and grow. I did yoga, and meditation, ate a clean, macrobiotic diet, and had early nights and lots of love in my life. I left where I was working, and started my own massage business in Bondi. Finally, I felt ready.

We thawed out two embryos. It was beautiful looking through the microscope and saying 'hello' to them before they put them inside me. The clinic had angels painted on the ceiling and played classical music during the transfer to make it as relaxing and comforting as possible.

When I came out of the clinic, I bashed my car into a pole in the car park. I thought, *That's so not me. Something's changed; I must be pregnant.* Two days later, Jonathon said, 'You're like a hot-water bottle. Your body temperature has gone through the roof.' We had a little chuckle. We both knew I was pregnant. We were so happy. We did a batch of home pregnancy tests and they all read, 'Yes, yes, yes!'

At my 37th birthday party at home, we announced we were having a baby. It was wonderful. There was so much happiness, enthusiasm, and joy because everyone knew the roller-coaster journey I'd been on over the years.

My first-time ultrasound showed two heartbeats and I thought, *I don't know if I want twins*. But Jonathon was over the moon. However, one of the heartbeats was much weaker than the other, and the specialist said, 'We'd like you to come back in a week and we'll do another ultrasound.' During that week the fainter heartbeat just faded off the scale. I was relieved, but Jonathon was disappointed. I viewed the other heartbeat as a little chaperone for the stronger one. It was his little pal to take him to where he needed to be, until he landed and then said, 'See you. You're safe and happy now; it's time for me to go.' I had a beautiful pregnancy and really enjoyed it.

I'd held a lot of babies, but I'd never held a newborn. Seeing Gus for the first time was surreal. He was so little and slippery, yet so strong. They had to break my waters just before the second stage, so he still had the slimy bag on him and was glistening. He was also very loud. I was exhausted in the first stage of labour and then I couldn't believe the surge of energy I got after the second stage. Wow, I'm alive again, and here he is! It was one of those moments where you think, *Far out! This is what being alive is all about. This is an amazing rite of passage, and I feel privileged to have been able to do it. Bring it on! I want to do that all over again! As soon as possible!* Because it was just wonderful. I loved it.

Sadly, Jonathon and I split up when Gus was 18 months old. There had been some serious problems in our relationship that I'd turned a blind eye to while I was focused on Gus. When I look back now at the day I went to the clinic to have the embryos transplanted and fell pregnant, it's symbolic that I went alone. Perhaps that was the beginning of the end.

Our son is called Augustus because he was conceived in August, although he was frozen and then transferred in January. Everyone says, 'Was he conceived in August? He can't be, if he was born in October!' That's our little secret. Now I've got nine frozen embryos, which I continue paying rent on — $150 for six months, the cheapest rent in Sydney.

I have fantasised about having one of the frozen embryos implanted in me, but now I'm single that's not really a viable option. It's a shame because we conceived those embryos when I was 36, and now I'm forty. I might not be as lucky next time if I do IVF with another partner. But who knows? In the meantime, I'm holding on to those frozen embryos.

I have girlfriends who are in their mid to late thirties and don't have partners, and I say to them jokingly, 'You can have one of mine! Sure, why not? Have one of our kids!' It's a tease to have all those embryos, because they represent possibilities and are part of the whole wondrous picture that created Gus.

After having Gus, I decided I wanted to become a birth-support person to help empower other women during childbirth. During my pregnancy, I didn't tell people how I'd conceived. But now it's something I often share in birth classes. You never know who's in the room and what experience they've gone through to fall pregnant. Maybe they'll be able to relate to my story and feel relief to know somebody else who's been through it, too.

I will tell Gus he's an IVF baby when he's old enough to grasp the concept. I think he needs to know that. Just as much as someone needs to know they're adopted.

Some of my friends still don't know he's an IVF baby. During the process I felt the need to protect myself from lots of people ringing and asking questions, or giving advice and their opinions.

When I was first pregnant, I was studying naturopathy and I remember my anatomy and physiology teacher criticising IVF: 'We should stop playing God and leave people alone who can't have babies, because it's fucking around with the natural population process.' I sat there seething, but then realised he was simply ignorant. So I kept it to myself. But the more I'm growing, and the more Gus is growing, the happier I am to talk about IVF. I don't see it as a negative thing — quite the contrary.

I would say to anyone considering IVF to focus on their pre-conception care. It's important for both partners to have healthy minds and bodies before conceiving. Sperm counts in the success rate of pregnancy, too. Good food and regular exercise benefits you ten-fold on a spiritual, emotional, and mental level to deal with the IVF process. I'd also advise women to see professionals in the field, not just take pot luck and read a few books. See a naturopath who specialises in women's health and IVF treatments. It's also important to embrace the treatment with as much positive attitude as you can muster. Don't resent or fear it, because it's hopefully going to bring you a beautiful baby.

CHAPTER 6
Desperately Seeking Daddy
–Gayle's story

Halfway through her second round of IVF, Gayle's partner walked out on her — leaving her, at the age of 40, single and desperate for a family. Determined not to compromise her dreams, Gayle decided that if technology could help her have a baby, it could help her find a partner as well.

I have one older sister and a brother ten years younger than me. There was another child after me who died soon after birth. As was the practice in those days, the nurses whisked the baby away as soon as it was born, so my mother never actually saw her little girl. My parents named her Diane, and she lived for only three days. There was no funeral or memorial service, and no proper grieving process for my parents. My mother rarely spoke about her.

I always knew that one day I'd have a child, but not while I was young. I felt I had too much to live for and too much to do. When I turned thirty, I said, 'Hurray, I'm still single and having fun and enjoying myself!' I had no concept about the possible difficulties associated with conceiving later in life. Now, fortunately, there's a lot more education and understanding about late babies. But at that stage, I thought I'd have one by the time I was 40 and it wouldn't be a problem.

I've been in financial services since I finished university. I worked for a company in Melbourne for fifteen years. It was later taken over by a bigger firm and I was transferred to Sydney, where I now live.

I was always a party girl. I never seriously thought about hitching up with any of my boyfriends. Probably the longest relationship was two or three years, but they never quite felt like long-term propositions. I certainly wasn't interested in having children with them. I was focused on having fun. I probably mucked around with the wrong types of blokes, knowing they weren't in it for the long-term either. I'd usually stay for twelve months or so before moving on.

When I was 35, I met a guy ten years younger and moved in with him. I'd just moved up from Melbourne and I didn't know anyone except him, so it was convenient for both of us. After we'd been together a while, I said to him, 'I want to have a child'. I think I realised by then my biological clock was ticking and I had to make some serious decisions in my life. He was relatively supportive, but wasn't prepared to make a commitment and get married. Deep down, I always knew he wasn't the right one for me, but I went off the pill anyway and we started trying to have a baby.

We tried for about six months before we sought help. We took natural remedies and different herbal medicines to increase our fertility, but nothing seemed to help. Finally, I went to a gynaecologist for a laparoscopy. He discovered fibroids in my uterus, which were blocking my tubes and preventing conception. They removed the fibroids and we tried again for a couple of months, but still nothing happened. My doctor, Rick, said there was a chance I'd have to do IVF because my partner didn't have the greatest sperm and I was getting older.

By then I was pretty determined to have a child. We did one cycle of IVF, and that didn't work. When we were halfway through the second cycle I felt intuitively I didn't have my partner's full support. That was confirmed the day he said he was relieved the first cycle hadn't worked.

'I just can't see myself wheeling a pram,' he said. 'I'm too young.'

Although he claimed to love me, he wasn't at the same place in his life, and we decided to spend a week apart considering our futures. During that time, I came to the decision that I definitely wanted a family and therefore couldn't stay with him. He came back saying he wanted out of our relationship, but that I should continue the IVF cycle with his sperm — a sort of farewell gift! But for me that just wasn't an option. I wanted to have a loving family, not be a single mum.

It was an amicable split until we came to the money side of things.

He still had his business set up in our house, which took a few months to move. Even though it was devastating at the time, I thank my lucky stars I didn't have a child with him. It wasn't meant to be and it wasn't going to work anyway. We'd only done one cycle, and by pushing him to do another cycle I'd tipped the relationship over the edge.

I remember going into the clinic and telling them I had to pull out of the cycle. It was hard because I knew I had little eggs maturing inside me and I had to let them go. I asked if they could harvest the eggs and freeze them, but apparently unfertilised eggs don't freeze very well.

I went to see a counsellor at the IVF clinic and she said, 'Your chances of having a child are very, very slim. You're 40 years old. Not only do you have to find another partner, you've got to go through all this IVF treatment again and then successfully have a child. Your chances are minimal.' I'll never forget that day.

If it had been me giving advice, I'd say, 'If you've got the right frame of mind and commitment, you can do it.' Even some guidance about how to get my life back on track would have been helpful, but there was none of that. Instead, she referred me to a psychologist. I went to one session with her and she said, 'Why don't you try and patch things up with your ex? Go and talk to him.' But I knew the relationship was well and truly over. He was ten years younger than me and we just didn't want the same things at the same time. I heard on the grapevine that he got married about a year later.

A few months later a male friend offered to be a sperm donor, but I really wanted to have a child with a partner. It wasn't just about the baby. I wanted to find someone to love and share my life with. Even my mother said, 'If you want to have a child on your own, we'll support you.'

Despite my friends' and family's doubts, I became even more determined to find a man who was at the same place in his life as me.

I believe sometimes you need to do something radical to get your life back on track. So about a month after the split I got onto internet dating. I went to a dating agency, and a woman helped me set up my profile with photos on the net. She helped me write an ad focusing on the importance of finding a man who wanted to have a family immediately. The only thing I lied about was my age; I said I was 35. About fifteen guys responded, and we contacted a few. One of them was Craig.

When I first saw Craig, I thought, *Well, he's not exactly my dashing Romeo with a six-pack and athletic build*. But he was tall and had a sense of humour, which were on my wish list. By the second date I knew he was a really nice guy and I could fall in love with him. We had some fun dates. We played tennis, and went to the theatre and art galleries. One day he cooked lunch and we drove to Palm Beach. Early on we started talking about having children. Craig was committed to having children by the time he was 40, and when I met him he was 37, just a few years younger than me. He likes to tease me now about lying about my age, but then I remind him his photo on the net showed him with a lot more hair than he has now!

When we started going out together, he said, 'Are you on any contraception?'

'No.'

'Well, shouldn't you be?'

'Look, the chances of me falling pregnant are so slim, there's no point worrying about it,' I said. I don't think Craig was quite ready to have children straight away, but it didn't stop us trying.

After we'd been going out for six months, I took Craig to see Rick, my gynaecologist at IVF Australia. I felt awkward because I'd been there before with another partner. I worried that it must have looked like I was desperate. It was a bit uncomfortable for Craig, too, because my previous partner's name was all over our file and they actually called Craig the wrong name a couple of times. Craig wondered whether my priority was to meet a man or have a baby. They certainly came very close. He jokes about it even now.

Craig's sperm tests turned out to have a high level of abnormal swimmers. He'd spent thirteen years in the Navy, and had been exposed to high-frequency and low-frequency radar emissions. Our gynaecologist said our chances of conceiving a child naturally weren't good.

Eight months after we met we did our first IVF cycle. The first one was hard because we had such high expectations. I seriously thought I was pregnant because my period was overdue and normally I'm very punctual. Craig wasn't living with me yet, and I remember being too excited to sleep. I sprang out of bed in the morning to do a pregnancy test and rang the clinic, and said, 'I've just done a pregnancy test, and there's only one line, but I think if it was really negative the line would have disappeared after a while and it hasn't.'

I was in denial and the clinic diplomatically advised me to wait a few days and come in for a blood test. I did and it was negative. My period came later that day. I cried all day and, when Craig took me to a lovely French restaurant that evening, I cried through dinner. Craig was also upset and disappointed. He bought a good bottle of red wine to drown our sorrows. He gave me a teddy bear called Honey, which still sits on my bed.

That night we had a serious conversation about whether we needed to consider donor sperm or eggs. It was tricky because we weren't sure where the problem lay. Over the years, a lot of my girlfriends had fallen pregnant accidentally and had abortions. I never had. So I wasn't sure whether it was me or whether it was him. Both men I'd tried to get pregnant with didn't have good sperm. I thought it would be easier and faster to find a sperm rather than an egg donor. Because of my age, I was worried we were running out of time. We also discussed using our brothers and sisters as donors, and we'd had some very generous offers from friends. However, out of sheer determination and stubbornness, we decided to keep trying with our own DNA while there was still a chance.

By the time I was 43 we had done a succession of IVF cycles, one after another. I could cope with failures if I had my mind set on starting a new one. In each cycle I'd never get more than five or six eggs. We used ICSI to inject the sperm into the eggs, and then only two or three would fertilise. The lab grew them to the five-day blastocyst stage before they transferred them into me. After the third cycle, Rick suggested implanting the blastocysts after three days. He said he'd had more success using that technique with older women. We were confident about our embryos because I was still producing reasonably good-quality eggs, which isn't always the case with women my age.

Craig said my emotions were up and down with the drugs, but I felt all right. I was on a high dosage, and hated using the nasal spray because it made me sneeze. I always injected myself, which was nerve-racking at first. But I thought, *I've just got to do this. I can't go to the clinic every morning to have injections.* I quickly got into the routine of injecting every morning, taking drugs, going in for blood tests, and having scans. I always went to the 7:00 am appointments so I'd be at work by eight o'clock, and I was never late. I only took a day off when I went to the hospital for the egg collection. My boss didn't know I was doing IVF. I didn't feel comfortable telling him because I didn't want it to influence my career.

Sometimes I felt tired, depending on what time of the cycle it was. But I just kept going and going and, as long as I was moving forward, I was happy. It would have been harder for me to pull out than to keep on trying.

It was always exciting when I started a cycle and had a scan to see how many eggs I had. Then, at the egg collection, I was buoyant if I got nine, and flat if I only got two. The daily phone calls from the scientists overseeing the fertilisation were nerve-racking. They'd say, 'Oh, sorry, only one's fertilised,' or 'Well done, five have fertilised!' Sometimes the embryos would die before the transfer date and I'd end up with nothing. But on the occasions I did have an embryo transfer, I'd think, *I've got a baby inside me*. I was so hopeful it was still alive and implanting. Then, as it got closer to when my period was due, I'd feel the cramping and dread the sign of blood.

I could cope with the physical side: the injections, hospital visits, having the extractions, the transfers, everything. But it's the emotional ups and downs that are hardest, especially as Craig and I did back-to-back cycles. I was in a mad panic because of my age. In hindsight, it was probably wrong to put so much pressure on my body.

We decided to have a six-month break and, ironically, I fell pregnant naturally. Craig proposed to me the night I told him. I didn't know he was going to propose and he didn't know I was pregnant. It was his 39th birthday. He'd always said he wanted to have a baby by the time he was forty. We went to a restaurant and I gave him his presents, and then I gave him a card to say we were expecting a baby. He proposed to me later that evening. It was a magical night and we were so jubilant.

We flew to Fiji the next day for a week's holiday. Before we left, I rang Rick. 'I'm pregnant!'

'Oh, see, you didn't need science after all,' he said.

But I started bleeding towards the end of my week in Fiji. I knew instinctively something was wrong when I saw the first spot of blood. The bleeding continued and I dreaded going to the toilet. The next day we went to a bird sanctuary, and I could have been on the moon. I felt so utterly numb. I was scared to admit our happiness and hope were over. On the way home, I sat in the Qantas lounge, feeling like the world was ending. I'll always wonder, *Should I have gone to Fiji? Was it my fault? Did something happen because I was travelling?*

Obviously the embryo wasn't strong enough to keep growing. When I got back home, I was still bleeding; Rick did a scan and there was no heartbeat, which you can see from about seven weeks. We left it another week, and I remember I had to do an exam for work. After the exam I took myself off to have another scan: there was still no heartbeat, and I was given a curette. Then I went back to work, which at times was my saviour. You have to put on a brave face at work; there's no room for pity when you're meeting clients and making financial decisions. I told myself, 'You must keep going and get back on the bandwagon. It's the only way.'

Then, on my next IVF cycle, I fell pregnant again, but I didn't hold it long — about seven or eight weeks. I was at work when I started bleeding. I came to hate the toilets at work because so often I'd discovered my period while in there, which meant any hope of a pregnancy was over for that cycle. I'd see the blood and think, *Oh no, not again, not again.*

Craig never told me to give up. He said, 'It's you who has to go through it so, if you want to, I'll be there with you.' He never whinged about having to do his little part in the process. We joked about which men's magazines the clinic would offer that month. He was always supportive. He could have said, 'No, we've had enough.' That sort of attitude puts so much pressure on you.

However, when Craig and I got to our tenth cycle, we both said, 'Okay, this is it, this is our last shot.' We did this special 'Colorado' thing where you do anything that might help. We took antibiotics and different sorts of steroids and vitamins. We threw everything at it. But it didn't work, and that's when we thought, *We've got to change our direction*. That's when we put an ad in the paper seeking donor eggs.

If it was the only chance, I was prepared to use a donor. It would still be our child. I've got a couple of girlfriends who have children from egg donors, and you would never know — not in a million years. One woman absolutely shocked me because I'd known her child for five years and had no idea she was from her sister's egg. Her second child was from a complete stranger. She even offered us two frozen embryos they hadn't used. It was nice to know we had a backup. My best friend offered me her eggs, but she was the same age as me, and the doctors weren't keen. We had offers of donor sperm, too, but we decided to use Craig's sperm and a donor egg.

Unfortunately, we didn't get many responses to our ad, and the few offers we had petered out by the time they'd met with the counsellor and the doctor.

Around the same time, my friend, who was pregnant using a donor egg, said to me, 'Did you know the government's brought in the safety net for IVF, and you only have to pay $800 a year, and everything else you get back on Medicare?' I thought, *Well, maybe we'll just do one more cycle using our own eggs and sperm while we're waiting for an egg donor*. When we told Rick we were going to try one more cycle, he said, 'Okay, I'll fill in the forms for you. But just because it's cheaper now doesn't mean it's going to work.'

But it did! On our eleventh cycle I fell pregnant. We couldn't believe it. We tried to stay relaxed about it, but we were both very nervous. It was a long weekend when I went for my egg collection, so I didn't have to take a day off work. And then I went away with my mum and dad for a week to Queensland. My period was due on the Monday, and I just had this feeling I was pregnant because I was so bloated. I waited until I got home on the Sunday morning to do a pregnancy test, and it was positive. We went straight into the clinic, and they did a blood test. They rang me on the Monday and said, 'Yep, you're pregnant.' But the pregnancy hormone was really low. The nurse said, 'Come back in a few days and have another blood test.' I went in on the Thursday and then went home to wait for their phone call. She rang and said, 'It's a really good reading.'

I was so relieved. But at the same time I was nervous. I was more positive than I had been with the other two pregnancies, but still cautious. When I went in for the seven-week scan, even Rick was surprised. 'Why on earth is this one working when the other ones haven't?' he said. 'Let's go and check if the heartbeat's there.' When we saw the heartbeat I started crying.

At ten weeks, I was in Melbourne for the birth of my girlfriend's baby when I started to bleed. I woke up in the middle of the night with some crampy feelings, and walked up and down the hall crying and saying to Craig, 'I can't lose this one, I can't lose this one! This is it.'

That weekend was hard for me because my girlfriend was giving birth and I thought I was losing my baby. It is hard when you see girlfriends who fall pregnant without even trying. This particular friend was using the withdrawal method when she fell pregnant. But I thought, *If it can work for them, there's hope for me, because we are all around the same age.* I tried to see the positive side.

When I got back from Melbourne, I rang Rick and he said, 'It should be all right if it's just a browny discharge.' And three days later I rang Craig at work and said, 'The bleeding's stopped. It's stopped!' That was just fantastic.

Rick suggested I have an amniocentesis because I was forty-three. It's a bit risky, but I wasn't so much worried about having a miscarriage as I was about the results. They showed us a chart, and at my age we were almost off the page with the risk of having Down's syndrome and other chromosomal deficiencies. After all we'd been through, we wondered what we'd do if the baby had Down's syndrome. We probably would have terminated it.

I cried during the test because it was the first time I'd seen the baby on the ultrasound, and it was amazing. We had to wait two weeks for the test results, which was really hard. When they rang and gave us the all-clear I burst out crying and rang Craig. I was blubbering so much he thought something was wrong. When the girls at work saw me, they guessed what had happened even though I hadn't told anyone I was pregnant. I was twenty weeks by then.

I didn't start to relax until I was about 30 weeks into the pregnancy. I panicked at every scan, worried they'd find something wrong. I held Craig's hand so tightly he thought it would break.

Despite the anxiety, I loved being pregnant. Craig and I walked every night after work to the new house we were building. I got a bit tired, but continued working until four days before I gave birth. I only put on six kilos and I had good blood pressure. I just loved it. The pregnancy went so well.

Because I'd had an operation on my uterus, they wanted me to have a caesarean. It almost felt like I was going in for another egg collection, because Rick was there with his needles and clinical equipment. And then it hit me. I wasn't going in for another procedure this time — I was actually having a baby! Even at that point, I was still scared something could go wrong. Then it all happened so quickly. Tessa was out and whisked away, and I was wheeled off to recovery. I felt I missed out on the early bonding we would have had if I'd given birth naturally and kept her with me.

I had problems breastfeeding because my milk took a while to come in. But we got through that trial and I breastfed Tessa until she was fifteen months old, which helped us bond.

Craig is the proudest dad. We just can't believe Tessa is our child. When I was pregnant, it was fun to wonder who she'd look like. When she was first born she looked just like Craig; now she looks more like me. It's nice to know she's our biological child.

We went back to see Rick after Tessa was born for a social visit. 'I shouldn't ask you this, but are you going to try for a second?' he said. 'I can tell you that your chances would be very slim, because by the time you'd be ready for another go, you'd be forty-five.'

'Don't worry, I doubt we'll be coming back for a second,' I assured him.

'You should count your blessings every day when you look at Tessa,' he said.

Rick was just fantastic. If it weren't for him we probably wouldn't have Tessa. Sometimes I'd go in there and he'd say, 'We can do it. We can do it.' And when I had my first miscarriage, he said, 'We'll get you pregnant one day.' Other times I'd go in there and he'd say, 'You're getting older and you might not be able to do it,' but I always felt he had confidence in me and appreciated my absolute determination. I was definitely one of his success stories. He was very proud of us. Whenever newspaper reporters rang to interview an IVF client, Rick always put them onto us.

Since we've been featured in a number of newspaper articles I've had many people come up to me and say, 'Oh, we've been told our only chance is IVF, but we're not sure if we want to do it,' and I say, 'Give it a go!'

Some people only do one or two cycles and they can't cope. Some women don't cope well with the drugs. Some don't like injecting themselves. Some don't like the emotion of it all. One woman said to me, 'I went through one cycle and it was all too emotional. I'm not going to do another cycle; it's obviously not meant to be.'

I suppose people have different tolerances, but I was never prepared to give it up. I had absolute determination. The injections were nothing, compared to having a child. I focused on the bigger picture.

I'll tell Tessa she's IVF. I'll also tell her not to leave it too late to have a baby and focus on finding the right man earlier in life — not to leave it until she's 40, like me. I'd like to see Tessa find her partner by the time she's thirty.

I was pregnant at our wedding, and Craig thanked IVF Australia for our baby in his wedding speech. I thanked some of my closest girlfriends, who have also had IVF babies, for their enduring support. I speak to the media and do whatever I can do to educate others about IVF. Ideally, I'd love to incorporate a career in IVF with the experiences I've had. It would certainly be rewarding.

Postscript

Tessa is eighteen months old now and absolutely adorable. And, believe it or not, we are going through a second cycle of IVF for another baby!

Rick couldn't believe it when we walked through the door and said we wanted to give it another shot. To his credit, he has supported our need to at least try for a sibling for Tessa. On our first cycle we only got one good egg, which fertilised but didn't take. I am on the highest dose of drugs possible to stimulate my ovaries, but it has been slow going. This cycle we have four eggs and the collection is next week. Fingers crossed.

CHAPTER 7

Defying Mother Nature
–Trisha's story

For Trisha, IVF was the easiest part of parenthood. Losing one of her twins and having the other born at 25 weeks were ordeals that left physical as well as emotional scars — so much so that Trisha believes it was her fear of another premature baby that stopped her getting pregnant a second time, despite ten more rounds of IVF.

When I was 22 years old, my appendix perforated and damaged my fallopian tubes, although I didn't know it at the time. When I was 26, I started getting terrible abdominal pains and abnormal periods. I was living in Queensland, and I went down to Melbourne for my niece's first Holy Communion. While I was there, my sister, who's a midwife, recommended I see an obstetrician she knew. He examined me and said, 'You've got a lump. I don't know what it is, but I need to investigate.' I said, 'Oh, fine. I'll book in.'

'No, no. You need to have it looked at right now.' His sense of urgency scared me. 'I want you to come into the hospital this afternoon. You should be in and out within an hour.'

I went in for the examination, and when I woke up from the anaesthetic and looked at the clock it was one o'clock in the morning. I had been there for ten hours!

'Why is it so late? What's happened?'

'Don't worry, go back to sleep. I'll come and see you first thing in the morning,' the doctor said.

When I woke later that morning, he told me he'd removed my fallopian tube because of the scar tissue from my appendix operation. The scarring was so extensive my whole reproductive system had adhered itself to the pelvic cavity.

I was in complete shock. 'But how can that happen?'

'It's not uncommon,' he said. 'Unfortunately, you'll need to come back every year to have further adhesions lasered off.'

But then came the worst news: 'You will probably never have children.'

I was incredulous. I'd always wanted lots of children, preferably six. So for someone to say I couldn't have babies simply didn't compute. I thought, *Oh, that's ridiculous. I will have children. He doesn't know what he's talking about!*

But then I was in shock. I couldn't breathe, and had dreadful anxiety attacks. The next day, I went to my niece's communion. Seeing those gorgeous little girls walk up the aisle in white dresses brought on another anxiety attack, and I had to leave the church.

But another part of me was strong and determined. I thought, *There's just no way. I* will *have a child*. I can't explain why, but I had a very strong feeling I would get pregnant.

In the meantime I went overseas to work, then came back and moved to Melbourne. When I was 30 I met my husband, Simon. I told him early on that I would have difficulty having a child, but he said it didn't bother him.

After we married, Simon was posted to London, so we moved to the UK in 1996. One day I was watching TV and saw a news item about a new IVF clinic that had just opened in Chelsea. I immediately contacted the clinic and asked if I could get on their waiting list.

Their first question was, 'Are you a British citizen?'

'No, I'm an Australian.' Fortunately, through patriality we had citizenship and were allowed on the IVF program.

I was excited, but Simon was really nervous. He couldn't even watch the injections without fainting. It did seem freaky injecting myself in the stomach that first time at home. But I soon got the hang of it, and didn't have any side effects from the drugs. For me, IVF was pretty easy and worked the first time. I knew I was pregnant almost immediately after they transferred two embryos into me. My sense of smell was heightened, and I was especially aware of London's disgusting diesel fumes and pollution. I did a home pregnancy-test and the lines went blue straight away. Because my hormone levels were so high we knew both embryos had implanted.

I rang my sister, Lorraine, and said, 'Guess what? I'm having twins!' She was beside herself.

'Give up work,' she said. 'You have to give up work!' She was so scared for me because, being a midwife, she knew how risky multiple pregnancies could be. She became increasingly worried when I developed hyperemesis and couldn't stop vomiting morning, noon, and night. I couldn't even keep down water. I was carrying way too much fluid, and felt absolutely massive.

My obstetrician did a urine test and found all my electrolytes were shot. I was really dehydrated, so he admitted me into hospital for a week. I remember feeling such relief when they put the drip in my arm: *Thank God I don't have to put anything into my mouth!*

I was very worried about the babies, but the doctors said they'd be fine. They were more concerned about me.

Then at seventeen weeks I had a spot of blood and rang my obstetrician at ten o'clock at night. He said, 'You'd better come in to the hospital and up to the labour ward. I'll organise for one of the obstetricians to examine you. You need to pack an overnight bag.' I thought, *Oh, surely not; it's only a blood spot.*

I was a bit scared, but Simon wasn't concerned at all. 'Don't panic, it'll be fine. It's only one spot,' he said. However, when we got to the hospital the obstetrician examined me and said, 'I'm sorry to tell you this, but your cervix has opened, which means you're going to go into labour, and you'll lose the babies.'

Because I was only seventeen weeks it was classed as a miscarriage, so I was put on the gynaecological ward, which meant Simon couldn't stay with me. They told him to go home, which added to my distress. Simon was beside himself, because I'd have to go through labour by myself.

But I was determined I wasn't going to lose anything and made Simon promise he'd come back as soon as it was light. He came back at 5.30 in the morning and demanded I be moved to a ward where he could be with me.

My obstetrician realised if I elevated my legs my cervix closed considerably, so he decided to try stitching my cervix shut. I went down to theatre and they sutured me and, as they were putting in the last stitch, they accidentally burst the first twin's sac, so they had to undo the sutures. When I woke up, they said, 'We're really sorry, but we've burst the sac around your first twin, which means you'll go into labour.'

I was hysterical and thought, *I cannot believe this is happening to me!* Despite that, I still felt determined to save my babies. They moved me to what they call the 'grieving room', which is a private room on the labour ward, because they thought I would deliver within 24 hours. But I didn't. They said, 'Okay, you can stay there rather than going back to the gynaecological ward.' And so I sat there until 21 weeks, when our daughter Harriet was born vaginally.

Harriet died straight away because her lungs weren't developed enough to survive. The doctors expected me to give birth to the second one immediately as well. But I didn't, and I'm not sure I can explain why, except I believe women have the most amazing power at times. I'm a very emotional person, and I knew if I cried or lost it during labour, I would lose the second twin. So I removed myself emotionally from the process and stopped myself going into labour. I managed to stay calm for the whole month, because I knew if I felt strong emotion, it would trigger the contractions and I'd lose the second baby. I refused to see Harriet, because I knew I'd break down if I saw or held her.

In the operating theatre they tied her umbilical cord and put it back inside me, but they'd never done that before, so they were very nervous it would get infected and I might die. They put me on a massive amount of antibiotics and watched me really closely.

During all this, Simon was still at work, and as soon as he arrived the doctor said, 'If your wife's temperature goes up, I'm afraid we'll have to terminate the pregnancy. Trisha could get septicaemia really quickly.'

Still, I knew if I calmed myself down, I'd be okay. That meant blocking out other people's comments, including Simon's and my sisters', who were saying it was unhealthy I hadn't seen the baby. But I knew what I was doing. Simon was a wreck, thinking, *God, I've just lost a child and now they're saying I could lose Trish, too. Maybe it would be safer to terminate the pregnancy.*

I stayed in hospital for another four weeks and didn't get out of bed except to use the toilet and shower. I couldn't do anything else. I tried to read, but my mind would go blank. Simon bought in a TV for me, but I couldn't watch it. My main aim was to stay calm and save the baby.

When I was 25 weeks pregnant, Elizabeth was born by emergency caesarean. The day I went into labour, Simon and I were stressed. We'd had our first-ever fight. He went off to a work dinner and I started contractions. He got worried and said, 'I'll stay,' and I said, 'No, please just go,' because we'd had enough of each other by then.

After he left, the contractions got steadily worse, and I became nervous and knew this was it. By midnight, they were unbearable. I said to the nurse, 'Should I ring Simon and tell him to come over?'

'Yes, maybe you should.' And as she said that, Simon appeared at my bedroom door. 'What are you doing here?'

'I don't know. I got home and had this overwhelming feeling I had to come to the hospital to see you.'

They then rushed me down to theatre, because Lizzie was going into foetal distress. When I got there a nurse came to me and said, 'Your sister Lorraine is on the next flight to England.' I thought, *When did Simon have time to call my sister?* Meanwhile he was thinking, *When did Trisha have time to talk to her sister?*

Apparently, my other sister, Kathleen, had woken up early in Australia after a shocking nightmare. She was in a cold sweat and had an overpowering sense something was wrong with me. She rang Lorraine, who was on an early shift at St Vincent's Private Hospital in Melbourne, and said, 'Something's happened to Trisha.'

'Don't be silly,' Lorraine said. 'I spoke to her last night and she was fine.'

'No, something's wrong, and I haven't got her telephone number. You have to ring her.'

'Don't be ridiculous. She's fine.'

'No, no, Lorraine, please ring her. I've got a shocking feeling about her.'

Lorraine ran downstairs and, with a bunch of dollar coins, rang the hospital in London and they said, 'Just one moment, I'll transfer you to the Labour Ward.'

Of course, that freaked her out. Then the receptionist said, 'Sorry, she's been taken in for an emergency caesar, and she's out cold.'

Hearing that, Lorraine got on the next flight to the UK.

At the same time, a friend of ours in London also woke up and said to his wife, 'We have to go and see Trisha.' That was bizarre, because I wasn't particularly close to him. I mean, my sisters I can understand — but why him?

I had an emergency caesar, and Lizzie was rushed to

the Neonatal Intensive Care Unit (NICU). I didn't see her immediately, because I was unconscious and then became quite ill. When I saw her three days later I got such a shock that I actually fainted. Nothing can prepare you for what a 25-weeker looks like. She was just a bit bigger than my hand and looked like a skinned rabbit covered in tubes. It was frightening and overwhelming. The nurse said, 'Would you like to hold her?'

'No, I'll wait,' I said. 'Just leave her there.' I think the nurses were worried we weren't going to bond but I thought, I can touch and talk to her, but to lift her up with all those tubes everywhere must surely hurt her — I'd rather leave her in her crib and not stress her out. But, no, they were determined for me to have a hold. There's a photo of us holding Lizzie for the first time, and you can see I'm terrified.

Somehow, though, I thought she'd survive. Even Simon confessed later that when I went into theatre at seventeen weeks, he had a strong feeling we'd end up with just one child.

However, we were probably the only optimists. When we lost Harriet at 21 weeks, the neotologists said there was a high chance of Elizabeth having cerebral palsy, brain damage, or blindness. They gave us the option to not resuscitate her when she was born.

We both said that they had to resuscitate, no matter what. I never thought I would make that decision. I'd always said I could never live with a child with cerebral palsy, but when it's a matter of life and death and you're given that option, you say, 'It doesn't matter. You must resuscitate, and we have to go through whatever happens.' And so, with all that in the back of our minds, Elizabeth was born weighing just 750 grams.

At first she was a good colour and doing well. I was able to give Lizzie my expressed breast milk through a tube. I was lucky, because often when a child is born prematurely your breast milk dries up from the trauma. For me it was the opposite; it poured out of me like tears. No one could believe it. They had to get another fridge for me, because I had so much milk. I offered to give it to other mothers with premmie babies, but wasn't allowed because of all the drugs I'd been on while in hospital.

Very few of our friends or family saw Elizabeth at that stage. We preferred it that way. However, three good friends came in one day to see her, and it scared the living daylights out of them. They didn't know what to say, 'Gee, she's really nice,' they said. Their faces were ashen-white.

Simon and I were determined to see Lizzie through this. Sadly, some parents didn't come in to see their premature babies. In some cultures, the babies are left there alone to see if they survive.

I stayed in the hospital for a month after Lizzie was born. Fortunately, we didn't live very far away. I would go in at seven o'clock in the morning and stay there until ten at night. Sometimes Simon and I would do night duty, depending on who was on the ward, because we didn't trust some of the nurses. We would sit there all night watching her.

Things took a turn for the worse when Elizabeth got strep and staph infections from someone in the medical team not washing their hands properly. We were ready to murder someone. Not long after the infection, Lizzie's heart failed and we nearly lost her. At 36 weeks she had to go to the Royal Brompton hospital for heart surgery. She was so sick we were told she would not survive. When she was transferred back to Chelsea and Westminster Intensive Care her heart and breathing stopped. For the first 24 hours after surgery she was dependent on life support, and the doctors said as a result she could have brain damage. We were told if after a certain time she couldn't breathe on her own, they would switch off the machine. As a result of the oxygen she also developed retinopathy — where the retina comes away from the eye and often leads to blindness. But it corrected itself, as did all her major problems, which is truly amazing.

During that time I made many pacts with God, although I'm

not particularly religious.

Simon and I would come home from intensive care, and get on the internet to see what other intensive-care units around the world were doing. We'd go back the next day with this new information and say to the doctors, 'What about trying this method? What about this drug? What about ...?'

We felt so raw ... that's how it is when it comes down to life-and-death situations. I saw a lot of death during that time. I heard women giving birth to stillborn babies. I heard them go through labour and then the terrible, primal out-pouring of grief. It was just horrible. I also saw babies die in the NICU. I was in the same room standing near a crib where a baby died. To be that close to a mother's grief is shocking. As a mother, all your instincts are to keep your child alive, no matter what.

I met a fifteen-year-old girl in the NICU who was pregnant but hadn't told anyone. She said to her friend one day, 'I think I'm having appendicitis. I have to go to the hospital.' She was 28 weeks pregnant and gave birth to a little boy. The midwife had to ring her parents to say, 'Your daughter's just had a baby.' She was from a well-off Chinese family. Only her mother and grandmother would come to see her. Her father disowned her. She used to say to me, 'Dad won't even talk to me.' And I said, 'He will, eventually. Don't worry.' It was so sad. The child had the most horrendous complications. He was blind, and couldn't feed properly because he had something wrong with his stomach. When I saw him a year later at the NICU Christmas party, he had a bowel bag.

Over time we became very close to the NICU staff, including my obstetrician. Sometimes he used to come down, after doing theatre, and just sit with Elizabeth and us.

It wasn't until Elizabeth was close to term at 40 weeks that she turned the corner. When she was quite well the staff wanted her to go into a paediatric ward, but I had to fight tooth and nail to keep her in the Neonatal Intensive Care, because I was frightened of her getting another infection in an open ward with hundreds of people coming in and out each day. But they needed the bed and wanted her moved. I said, 'I don't care. She's staying here, and if you remove her I'll take her home.' They didn't want me to do that, so they had no choice but to keep her there.

Simon and I were very naïve when we walked into that hospital at seventeen weeks, but when we walked out six months later we were two very different people. We became really strong; we'd question everything and challenge the medical staff, which is something I would never normally do. But you learn to do that, because that's how your child survives. I'm no different to any other mother; you just do what you have to do.

Lizzie stayed as a neonate until we took her home. We set up our house in London like a hospital, with all the necessary equipment, including oxygen tanks. She stayed on oxygen for the first year of her life. In the first few months a nurse came every day to check on us. We heated the house to 29 degrees; it was like the tropics. No one was allowed to come in without washing their hands or taking off their shoes. I practically stayed in the house for a year.

Our goal was to get her well enough so we could bring her back to Australia, which we did, and we moved to Sydney.

Lizzie progressed well, but I was still very nervous in the early years. She used to get chronic tonsillitis every year. When she was four, she had twelve bouts in one year and the antibiotics didn't work for her anymore. Lizzie was admitted to hospital countless times and put on an antibiotic drip. One day when she was particularly ill, I rang the doctor and said, 'I don't know what to do. Should I take her straight to the Royal North Shore Hospital?' and he said, 'I'm just around the corner. I'll come and give her a penicillin shot.' He came and looked at her and said, 'Patricia, we've talked about this. The penicillin isn't working anymore. She's really, really ill. She's got to have her tonsils out.' I was so terrified.

'There's no point talking about it, because I'm not having her tonsils out. I'm just not doing it. I can't watch her go through another operation.'

'Why don't you want her tonsils out?'

'Because I'm scared she's going to die. Once you give me a 100 per cent guarantee she will be fine and her lungs will cope, I'll do it.'

'You know I can't do that,' he said. 'It's always a risk for a child to undergo a general anaesthetic. I can't force you to get her tonsils out, but Lizzie's really ill and there's a risk of septicaemia.' When he left, I began howling, and realised my fear was making her sicker rather than helping her. So I rang him back and said, 'Okay, let's go ahead.'

We saw a brilliant ENT specialist, who arranged for a special anaesthetist, and I told him my fears. Then Elizabeth's paediatrician rang and told him her history, and he was very understanding. 'Why didn't you tell me you'd lost a child?'

'Well, I just didn't think it was relevant, really,' I said.

Anyway, they did a fantastic job. Poor Lizzie's tonsils were so huge, they were impacted into her neck. Later the doctor said, 'It was the best thing we could have done.'

But then Lizzie wouldn't eat or drink for ten days, because it hurt her to swallow. I used an eyedropper to get fluid into her. On the tenth day I said, 'Lizzie, if you don't drink you're going back into hospital.' She was so tired of continually going to hospital and being put on a drip she said, 'No, I'll do it!' She drank, and that was a massive turning-point. Lizzie put on weight; and her hair, which used to be fluffy and wispy, became thick and lustrous. She obviously hadn't been absorbing nutrients. That was the turning-point, and we have not looked back.

Just when I thought my worries were over, Simon suggested we go back and try IVF again for another child. 'Oh, it would be really good for Lizzie to have a brother or sister,' he said. He was much keener than I was.

All I could think was that I couldn't go through surgery again. I couldn't go through that emotional turmoil of having another premature baby. I'd be in a loony bin. That experience with Lizzie had taken every ounce of emotion and fear out of me.

And yet I agreed to go ahead. I thought I was being unfair to Simon and Lizzie. Maybe Lizzie did need a sibling after all.

However, the whole time I was on IVF I was terrified I would get pregnant. I had so many fears. I'd had a classical caesar, and the doctor told me that the scarring on my uterus could perforate if I got pregnant. I was also scared of having twins again.

While I was doing IVF I put on a massive amount of weight, probably as protection or comfort. We did ten cycles of IVF back-to-back. We did three fresh cycles, and the rest were frozen. They put two or three embryos back in each time. I didn't get pregnant once. The IVF clinic doctors were very surprised, because I have one good fallopian tube and I have no problem producing eggs. My sisters are all very fertile, too. I think I subconsciously blocked it. Then I turned 40 and thought my chances were decreasing. That's when I finally said, 'It's not meant to be and I'm not continuing any more. I gave it my best shot, and it didn't happen.'

Simon was okay about it but said, 'How would you feel about adopting a little Chinese or Indian girl?' I told him he was mad, because there's such a long waiting list — at least two years. But we'll go through the process, and see how it pans out.

Sometimes I worry about the long-term effects of all the IVF drugs, but what can you do? When I had IVF with Lizzie back in the nineties, they said to me, 'We don't know what effect it will have on you. We can't guarantee you won't get cancer. We recommend you have breast screening and check-ups every year,' which I do anyway. But then, I had a friend who just died of motor neurone disease — and he was a non-smoker, a social drinker, and used to run every day. He was a healthy 42-year-old. So you just never know. You can also get pregnant naturally and die, too. Life's a gamble.

Apart from her asthma, Lizzie is now a normal, happy, healthy nine-year-old girl. I haven't told her yet that she's IVF; I think it's irrelevant. Being a premature baby had much more of an impact on her.

She's got a big scar on her back from her heart operation, and when she was at swimming lessons one of the other girls pointed and said, 'Oh, what's that huge scar on your back?' Lizzie had never noticed it before and I never thought to point it out to her. She came home really upset: 'Mummy, why have I got a big scar on my back?' I just felt sick.

'Well, Lizzie, that's from your heart operation. That's your lucky scar, just like Mummy's got a lucky scar where you came out of me,' I said. 'Everyone has scars, whether they have them now, or later on in life. It's no big deal.' She was fine with that. I think she was more upset at it being pointed out and teased.

I still think IVF is an amazing process. Those people are so gifted. It must be an incredible feeling to know you have been part of a process that helps thousands of women and men become parents.

Postscript

Trisha and Simon have just had their application approved to adopt a baby girl from China. They are now waiting to pick her up.

In the meantime, they took Lizzie to England to visit the hospital where she was born and cared for in those early critical months.

Some of the original staff were still there. They cried when they met Lizzie, who was quite overwhelmed by the visit.

CHAPTER 8

Sister Versus Sister
–Sandra's story

Sandra had spent her professional life fighting for social justice. But when her niece agreed to donate her eggs to her, she could never have predicted how the decision would rip her once tight-knit family apart and shatter her illusions about love, blood ties, and fairness.

I was one of four children. My mother's second child died of leukaemia. Mum never spoke about it, and there was only one photo of him in the house.

When I was growing up I never envisaged having children; it never entered my head. As an adult I've enjoyed being heavily involved in political and social-justice work. I had a couple of terminations when I was in my twenties — at the time, it seemed there was no way I could have had those children. My sister Bev, on the other hand, had five children, and I enjoyed being around them.

It wasn't until I met my husband, Paul, that I began to think about having children. He had one child already and, after he separated from his wife, his daughter came to live with us for a while. She was only five, and found it hard adjusting to the new situation. While I would have liked to start a family then, it wasn't the right timing. When I finally brought up the subject, Paul freaked and said he wanted us to travel instead. It became such an issue that we went to counselling, and resolved to travel first and have a baby later. However, when we came back from overseas my eldest brother was dying of cancer, so I delayed getting pregnant to spend time with him. It took me quite a while after my brother's death to focus again on having a child. We tried for about eighteen months, and I finally fell pregnant when I was forty.

Having my daughter was an overwhelmingly wonderful and thrilling experience. It was the best thing that had happened to me. Paul was delighted at becoming a father again, and is a fantastic dad. I found motherhood utterly consuming. I breastfed for three years. My daughter would wake every two hours during the night, and feed for long periods. She was very active and demanding. Those first few years were exhausting, but full of joy.

It wasn't until my daughter was three years old that I started to think it would be lovely for her to have a brother or sister.

I mentioned it to my GP and he said, 'You're going to have to try especially hard because of your age.' Being so sleep deprived, the last thing I felt like doing was having regular sex. Nevertheless, we tried for a while and then I thought, *This is crazy. I'm not getting any younger. I need expert advice.* So I went to a women's health centre in Canberra, and they eventually sent me to have blood tests to check my ovulation. To my surprise, the tests showed I was pre-menopausal. Looking for answers, I went to a fertility gynaecologist, and the first thing he said was, 'We'll get you pregnant, no worries.' However, when I told him I was 44, he did a complete turnabout. My initial surge of hope suddenly fell flat. Had I gone to him earlier, perhaps I might have had a better chance.

By this stage, my periods were coming every second or third month, and I was experiencing occasional hot flushes. For some reason I still had faith that I might fall pregnant. I said to myself, *You've just got to try, but don't get your hopes up*. Of course, I did get my hopes up, and by then wanted another child badly. Every time I looked at my daughter I had an aching feeling. She adored babies and would have loved a sibling. I thought another child would be a blessing for us all.

I started taking Clomid to stimulate ovulation, and went to a naturopath who specialised in fertility. I did everything I could think of. Eventually it became clear I wasn't ovulating. The gynaecologist said my only option was egg donation, which had never occurred to me and I knew very little about it. But as I learnt more, I thought, *Okay, it's not what I ideally wanted, but it's an option*. The doctor suggested finding a donor I knew, which was tricky because most of my friends were the same age as me, so their eggs wouldn't have been viable either. I racked my brains. *Who can I ask?* It's not exactly the sort of thing you can just ask anybody. Because of my age, there was pressure to find somebody quickly. Eventually I thought of my niece, Julia, who was 27 at the time.

We thought Julia would be perfect. However, a doctor warned that egg donors who haven't had their own children might feel overly attached to the child. For that reason, they prefer donors to have finished their families. The other concern was that if a complication arose during the egg harvesting it could potentially affect Julia's future fertility. Obviously I didn't want to put Julia or anyone else in any danger, but the IVF specialist assured me the risks were minuscule and there was no need for concern. The issue still remained, though, as to whether my niece would feel okay about it.

After a lot of soul searching, I finally asked Julia. She laughed and said, 'Of course I will! It's only an egg. I'm not giving you a lung, for goodness sake!'

I was relieved she was so supportive, but refused to take that as a definite 'yes'.

'I'm incredibly appreciative, but you need to be absolutely sure and check everything out. I'd like you and your husband to see the IVF specialist, and talk through all the medical and emotional issues with a counsellor.'

I gave Julia all the information I could find, including newspaper articles and medical journals. But she was so busy with work she wasn't interested in reading them, so I verbally summarised them for her as best I could.

Julia and I and our husbands underwent a series of compulsory counselling sessions at the IVF clinic. My niece's husband was also very supportive, despite it being completely new territory for him, too. The counsellor posed all sorts of issues and hypothetical scenarios, including my niece's potential feelings toward the baby.

'What if the baby looks like you? How would you feel?'

'I can't see it being an issue,' Julia said.

'Come on, imagine a gorgeous little baby girl who looks just like you. You wouldn't feel like her mother? '

'No, really, it wouldn't worry me. It wouldn't be my baby,' she insisted.

Then the counsellor suggested I speak to Bev, Julia's mother. I wanted to talk to her face to face because it was such a private matter. I decided to wait until we'd completed counselling and were positive we were going ahead. I turned up at her house a few times, but it wasn't convenient or private enough to talk. Unfortunately, by the time I did tell her, my niece had already begun taking the hormone drugs. A part of me thought my sister would be supportive and happy for me, but another part was worried about Bev's strong religious background, which might make it hard for her to accept the concept of egg donation.

I told Bev the whole story truthfully and honestly. I said we'd been trying to have another child for a long time, and the only option left was egg donation, and Julia had agreed to be my donor.

'Oh, she would, wouldn't she?' Bev said. 'Julia's experimental and would do almost anything because she doesn't think things through.'

She became highly upset and accused me of abusing her daughter, saying Julia wasn't emotionally mature enough to make a decision like this. Nothing could have prepared me for Bev's response. One of her strongest concerns was that her daughter didn't know what it was like to feel a bond with a child because she'd never had one. She was convinced that as soon as Julia saw the baby she'd want to keep it.

'Why couldn't you have asked one of Julia's friends instead? Then she wouldn't have to be confronted with seeing the baby all the time'

'We went through counselling,' I said. 'We covered that scenario, and Julia was adamant she doesn't want a child yet because she's focused on her career. The counsellor is content she's clear about that.'

'Oh, counsellors are just irresponsible!' Bev said. 'They're biased anyway because they're linked to the IVF clinic!'

She then became irrational, 'You're stealing my first grandchild!' she said. 'You'll tear the family apart by doing this.' I thought that wasn't remotely possible, because our family was so close. But, as it turned out, that's exactly what happened.

Finally I said, 'Okay. It can be stopped. It's not a fait accompli.'

I was completely shaken by Bev's reaction and went immediately to see Julia.

'Your mother is very upset. She's completely opposed to this and you need to talk to her,' I said.

At first my niece laughed and didn't take it seriously, but then she said, 'Don't worry, I'll talk to Mum.'

I was so cut up, not because I thought I was going to lose my egg donor, but because Bev had accused me of abusing my niece's 'naïvety', which I would never do in a million years. I believed I was very thorough about getting accurate information to her.

I decided to talk to my mother. 'Mum, I'm so upset, I don't know what to do.'

'Julia wants to help you because she's a family-oriented person and it's a lovely thing to do,' Mum said. 'If you had done that for somebody else, I'd be really proud of you.'

I burst into tears, and thought my mother was wonderful and so supportive. I told her I thought I should stop the IVF treatment.

'Don't worry, I'll talk to your sister,' she said. 'Don't stop the process. Go ahead.'

To have my mother's backing and reassurance meant a great deal to me. I never wanted a rift in the family; otherwise I would never have embarked upon it.

I still wondered if I should stop Julia taking the egg-stimulating drugs. But part of me thought, *I shouldn't be telling her what to do. It's her decision. She's old enough to make up her mind.* My head and heart were in conflict, so I left it for a while to give me time to think.

However, things took a turn for the worse. Bev phoned my husband the next morning while I was out and accused us of preying on my niece.

Before we'd had a chance to talk to the rest of the family, Bev had rung everyone, lobbying against what I was doing.

Bev spoke to my mother, who went from being supportive to dead against it. Bev claimed she'd spoken to a nurse who told her the egg donation would adversely affect Julia's future fertility. I knew the information was wrong, but my sister was on a mission. She constantly came up with all sorts of anti-donor ideas, some of them quite off-beam. In her view, she was on a campaign to protect her child from abusers.

It was very upsetting and I felt betrayed by my mother. She was suddenly so negative and saying, 'Yes, but what about this problem and this and this?' Although I tried to allay her fears with logical explanations, she had all this crazy stuff in her head from Bev. Her strongest concern was that Julia would be upset when she saw the child.

To this day, I don't believe that was likely. But the genetic-bonding issue became a really strong theme in Bev's attack on me. She was convinced if Julia saw the child regularly she'd covet it, especially if it resembled her. I thought donating her eggs might be something Julia would view positively, and she'd be happy if the child looked like her. And, in any case, we're family, so there's a resemblance amongst all of us.

My husband and I couldn't sleep at night. We were so utterly shocked by Bev's vitriol. She even sent a series of acidic letters which I refused to read but, Paul said, contained all sorts of strange attacks on me. One thing that really grated on my sister's nerves was that she wasn't told from the outset. She felt she should have been asked before my niece.

In a letter to my mother I explained, 'I would never have asked somebody who I didn't feel was absolutely capable. Julia is highly intelligent, academically qualified, and adept at seeking out information and understanding it.'

Bev read this letter and said, 'Oh, so you're saying I'm not knowledgeable. Is that it?'

I think, at the heart of it, Bev felt inferior and upset because she wasn't consulted before her daughter and that things were beyond her control.

She phoned me and raved and ranted, and I just listened to her. I didn't counter what she was saying, because I didn't think she'd hear it. She was so full of anger.

I wondered if she was jealous because I'd had a career and she'd missed out because she'd had children quite young. She was also long separated from her husband, and still harboured a lot of bitterness.

I remember, many years ago before I had my daughter, Bev mentioned a new scientific finding that claimed mothers who breastfed had less chance of getting breast cancer, and she made a mocking remark about childless career women.

I was quite taken aback. 'Why would you wish that upon anyone? Women who don't have children or breastfeed aren't evil!'

She had a completely different world-view to me, and wasn't able to look at another perspective. She made up her mind on the spot, the minute I told her what my niece was doing, without any consideration. She has very set ideas about genetics and love. She claims even adoptions cause dysfunctional families. It's true, they have caused problems in the past, but society has come a long way since then. Today, people are aware of the importance of talking to their child about who the egg or sperm donor is, and encouraging contact. And in our case it would have been ideal, because the child would have known from early on who its biological mother was. In my view, the child would have brought joy and love to the whole family. But not in Bev's book.

I had only one brother, who listened to me, and I spent many nights at his house in tears. He was upset, too, at the way I was being treated. When he got acrimonious phone calls from Bev, he said, 'Please stop this! It's not on.' He was fantastic. It was so important to have somebody stand up for me.

Eventually, my mother and sister's campaigning wore my niece down, and she rang me to discuss it.

'We've got to stop. You've got to stop the drugs and the whole process,' I said.

Julia agreed, and we actually both felt a huge sense of relief. But no one could have envisaged the long-term damage it would cause.

Christmas that year was traumatic. We've always gone to my family's house in Canberra, and I've loved being with them. But I wrote to my mother saying I wasn't sure I could be there. Then Bev rang me just days before Christmas and abused me about the letter. She was explosive and verbally violent. The strength of it was so phenomenal I had to hold the phone away from my ear.

So that year we went to my in-laws' for dinner. I rang to wish my family 'Merry Christmas', and left a message on the answering machine. No one returned my call, not even my mother. It still hurts to think about it.

During the Christmas break I saw a counsellor. I decided I would 'behave well,' and even if my mother and sister weren't repentant for their behaviour, I could forgive them. I offered an olive branch by writing to my sister, apologising for not telling her about the egg donation earlier and saying that I valued our relationship. Sadly, it didn't help, and her anger only grew.

It's more than a year ago now, and the relationship with Bev is still strained. She continues to send poison-pen letters, and is utterly cold toward me at family gatherings.

People said my niece was relieved it was all over, and glad her mother had stepped in to 'save' her. Later I spoke to her about it and, fortunately, we still have a close relationship. Thank God, that hasn't changed.

One of the worst consequences to come from of all this has been that my sister has cut off my daughter by not inviting us to family functions. My daughter will now grow up not only without a sibling, but also having limited contact with her cousins. She hears about family gatherings where she's not welcome, which is very cruel.

For me, my mother's response has been more hurtful than anything my sister said. When I asked her why she took Bev's side, she said it was because my sister was on her own. My opinion of my mother has changed. I've had to admit she doesn't love me unconditionally. Counselling has helped me put it into perspective. I now think, *Okay, that's who she is, I don't like it, but there's a whole lot more to our history than this unfortunate episode.* I try to look at her good points, and our relationship is fine as long as we don't discuss that topic.

After a time of grieving, I found a wonderful egg-donor through advertising. But, by then, I was so much older and battle weary that I didn't have it in me to continue. When I told my mother I was considering using another donor, she said, 'Yes, that'll be fine', but you could see the tension and anger in her from thinking about the past. And I thought, *Do I really want to bring a child into a family with all these negative views about donor-egg children?*

I've now come to accept I have one child and that's fine. I just need to ensure she has plenty of other children for company, and that we build on the strong relationships around us so she'll have people to depend on when Paul and I are gone.

Through all this I've learned it is possible to forgive others, regardless of how they treat you. It's important to take happiness into your own hands. I read Dickens' *A Christmas Carol* recently, and took on board the message to keep pouring love into those who reject it, in the hope that, one day, some of it enters their hearts.

CHAPTER 9

Double Life
–Lauren's story

Two husbands, two vasectomies, two rounds of IVF, and twins — welcome to Lauren's double life.

When I was growing up, IVF was still quite controversial. I gave a speech about it when I was in Year 11, arguing, 'Who are we to say whether it's right or wrong for someone to use technology to have their own precious baby? If the technology's there, why shouldn't we use it?' There were quite a few raised eyebrows in the audience. IVF was still relatively new then, and many people thought we shouldn't interfere with nature. By the time I did IVF, it was so commonplace that people would say, 'Oh, you've got twins. Which clinic did you go to?'

    I was 23 when I met Colin. It was on a day trip to the reef outside Townsville, and he was the dive instructor and took me on my first dive. He had bright blue eyes and a huge grin. There was immediate chemistry between us. He invited me to a party and was shocked when I said 'yes'. He was a lot of fun, and we soon became besotted with each other. Colin and that first dive made such an impression that, eighteen months later, I became a qualified instructor, too, and we started our own dive business on Magnetic Island.

Then, about six or seven years later, I was suddenly hit with the hormone stick and desperately wanted to have children. Colin was twenty years older than me and already had a family from a previous marriage. He never dreamt he'd have more children and had had a vasectomy. We realised we'd need to do IVF when the time was right. Our diving business was all consuming, so taking nine months off work wasn't an option then.

But when I was 30, I had treatment for endometriosis, and the gynaecologist said to me, 'If you're going to have children, do it now, rather than later.'

Thinking our issue was just a 'boy's problem', we went to see an urologist. I remember this tiny man behind a great big desk who sat us in chairs much lower than his, with our knees virtually under our chins. It was really quite funny. It had been fifteen years since Colin's vasectomy, and a reversal didn't appear viable. After listening to our story, the little man said, 'Your best bet is to go straight to the gynaecologist downstairs and organise donor sperm. You've got no chance otherwise.'

I felt such a goose when I said, 'Actually, I saw this technique on TV, where they extracted the sperm from the testes and injected it into the eggs in a petri dish. Could we try that?' I didn't have much information, and he was very disparaging that my sole medical source was *Oprah*. Anyway, we saw the doctor downstairs and he said, 'That's rubbish, you don't need donor sperm', and he gave us some information about alternatives. We contacted Monash IVF clinic in Brisbane and asked if they did ICCSI, the technique I'd seen on TV. They said, 'Yes, of course, we know all about it and use it here all the time.' So we decided to move to Brisbane and give it a go.

We started with treatment for my endometriosis. I had heavy periods, which sometimes made sex painful. Every month I'd miss a couple of days' work because of the pain. The doctor tried a couple of laparoscopies to laser the endometriosis, but it was in a difficult spot to reach. I ended up on drugs for six months that made me feel homicidal. My husband said it was like living with a rabid blue-heeler. He was afraid I'd bite off his hand if he looked at me the wrong way!

Once we started IVF, however, there was so much excitement and joy that it was finally happening that I didn't notice the side effects as much.

We committed ourselves wholeheartedly to the process. We gave up coffee and chocolate, and took vitamin supplements to make ourselves as healthy as possible. We saw it as preparing for a special mission. To us, IVF was a unique opportunity and not an ordeal. We entered it with a heart full of joy and not too much anxiety. I found it helpful talking to people about it, too. We were surrounded by people's good wishes and positive energy, which put us in a good mental state to tackle the clinical process.

The doctors had a few plans to get around Colin's vasectomy. The first was to go into the tubules in his testicles and hopefully find a few partially formed sperm. All they needed was the nucleus, not even a whole sperm. If that didn't work, they would take a piece of testicular tissue and hope they'd find a few in there. The last resort was to have donor sperm on standby.

I was upset by the prospect of having to use donor sperm. There was something about Col's beautiful blue eyes that melted me. I hadn't realised until then that part of my dream was to see those eyes in our child. The clinic gave me a catalogue of potential sperm donors. There were only two with blue eyes, and they sounded nothing like us. One had auburn hair and was much shorter than us, while the other had dark hair and was a giant. There was less information about them than in the personal columns, and somehow I was supposed to find a suitable biological father for my children. It listed their occupation and education level, but I wanted to know what made their hearts sing, whether they laughed at Gary Larson comics, whether they liked diving, or what they thought of *To Kill a Mockingbird*. It was touching that someone was willing to donate to make our dream come true, but it would have been a slightly different dream and I was surprised how much that mattered to me. Fortunately, in the end, we didn't have to go down the donor-sperm path.

Colin was quite good about it. Obviously, he wished we could conceive naturally and felt it was his fault we had to do this. I also wished it was easier and cheaper. On the other hand, Colin was thrilled to be given a chance to have another family. He'd never dreamt it was possible, and I was happy he was prepared to do it with me.

Colin gave me the daily injections because I was too squeamish. But on the night of the extra-large needle to trigger ovulation before egg pick-up, he wasn't home and I had to inject myself. I sat staring at the syringe for almost an hour, trying to summon up the courage. Once I finally did it, I wondered what the big deal had been.

We got thirteen eggs from the harvest and seven fertilised, which is pretty good. The night of the egg collection I dreamt I was woken by the noise of a rowdy party at the bottom of our paddock. I went to investigate, and stood at the door watching a crowd of people in medieval dress, laughing and dancing around a huge bonfire. There were horses and dogs there, too. When they saw me, a couple of people hugged and waved goodbye to their friends and then rode up to me at the house. Their friends continued their party by the bonfire, toasting them and waving farewell and good luck. I interpreted it to mean that the souls of our embryos were setting off for their new incarnation as our children. It seemed like a good omen.

Although the embryo implantation was quite a difficult transfer, it was also rather funny. I had my legs up in stirrups, and I remember watching what the doctor was doing in the reflection of his glasses. He had his tongue poking out the side of his mouth with concentration. The technicians and nurses kept wandering in and out, and at one stage there must have been eight people in the room. There I was displaying my wares, and someone was chatting about going on holiday. If I hadn't been semi-naked and in stirrups, it might have been a pleasant social gathering! It certainly wasn't the most romantic way to conceive.

Eventually when the two embryos were transferred, everyone left, and Colin and I had the room to ourselves. It was supposed to be a magic moment, but it was such a strange atmosphere that I couldn't relax.

We decided to donate our other five embryos. I couldn't bear the thought of them not having a chance, only I didn't fancy having seven children. It was a heart, not a head, decision. I wanted to share my joy. Having children was something I wanted so badly for so long, and with a flick of a pen I could make that possible for someone else, too. I don't think that is such a big ask.

But it turned out to be a sad story. We donated to a woman who had been trying IVF for ten years without any luck, even with donor eggs. She took three of our embryos and became pregnant after implanting one. She carried to twenty weeks and then the ultrasound revealed the foetus had multiple disabilities, including Down's syndrome, so she terminated it.

When I heard the news, I thought, *Oh, God, what does that mean for the other embryos — could they also have Down's syndrome?* But the clinic said, 'You have no claim on those embryos anymore. You've signed them away. We'll let you know if there's another conception.' I've never heard from the clinic, so I don't think she fell pregnant again.

Under Queensland law, the offspring of a donated embryo, egg, or sperm can contact you when he or she turns eighteen. During the required counselling session, before we agreed to donate, the counsellor asked, 'What are you going to tell your other children about their siblings if they turn up on your doorstep one day?'

That got me thinking; so over the years I've kept a folio of newspaper articles and information about embryo donation and IVF, which will hopefully make it easier to explain.

Also in the folder is a long letter I wrote my children the night before they were born. This is an extract:

> Then we had to wait two weeks to do a pregnancy test, to be really sure you'd stuck around. I felt mixed emotions, the same you get waiting in line for a roller coaster: desperate to do the test, but terrified of the answer. During the waiting period, we interpreted every little sign as an omen. The bird of paradise in our garden produced two beautiful flowers that week; it has never flowered before or since. We finally did the test at 1:00 am, and it was positive! We were thrilled and overawed by the enormity of this new knowledge and also a little intimidated. Then we got on the phone because there were lots of people who wanted to know, as soon as we knew, even at 2:00 am!
>
> The next day we did the official blood test at the clinic. We'd grown so close to the staff, they cried with us when the test came up positive. You need a blood HCG hormone level of more than 25 to be pregnant. My first test had a reading of 150, and two weeks later it was 2000 — so obviously there was more than one baby. I was thrilled to be having twins, but Col was horrified. I thought it was such a bonus. We'd waited so long, and now to be pregnant with two was to be doubly blessed.
>
> My pregnancy was mostly a dream run, although I had a complication from the IVF at nine weeks when my fallopian tube twisted. It's not very common but, if you've got longish tubes, they can twist from the added weight of having had so many stimulated follicles. I hadn't heard about it before, and it wasn't something we'd been warned about. One minute Colin was racing me to casualty, while I was puking, doubled up with pain and thinking I was losing the babies; then, the next, my fallopian tube righted itself, and I said, 'Oh, I'm okay now!' It was a strange situation — terrifying and then a little embarrassing. But we had another scan at the hospital which, fortunately, showed

everything was normal.

The rest of the pregnancy was fine up until 35 weeks, when I developed pre-eclampsia. They gave me steroid injections to prepare the babies for an early delivery, and our beautiful boy and girl were born via caesarean section at 36 weeks. We named them Hamish and Georgia.

When I first saw them I was very teary and happy. Hamish had difficulty breathing, which sometimes happens with premature babies. I only saw him for a moment before they took him away to the special-care nursery for 24 hours. I was horrified, but the nurses reassured me he'd be fine. I was pretty out of it on morphine and hardly able to keep my eyes open. I remember watching Georgia in amazement, and counting my blessings that she and Hamish were finally here. I only had a Polaroid photo of Hamish with me for comfort and it's probably just as well I was drugged to the eyeballs, otherwise I'd have been demented with anxiety. Fortunately, by the time I came around, Hamish was back with me. The sense of bliss at their arrival is still with me today. And wouldn't you know — Hamish has his father's beautiful blue eyes.

The first Christmas with the kids was very special. Ever since I'd started to desperately want children, Christmas had been a bittersweet time. It was awful to be around children and feel the terrible absence of my own. Every year I thought, *God, is this the last one without children?* When our first Christmas with the twins came, and despite feeling like a sentimental fool, I went down to the IVF clinic with a card and a photo of the babies. When I got to the reception, the counter was covered in hundreds of cards, all reading, 'Our first Christmas with our babies has been so special. Thank you!' It took my breath away.

Having twins, people immediately raise the question, 'Did you do IVF?' I'm used to it now and we have quite a few friends who've also had IVF twins, so it's the norm in our world. I don't feel like the IVF tag is anything to be ashamed of. But not everyone sees it that way. When I was pregnant I was working with a friend who had naturally conceived twins and when people would say to me, 'Oh, you're having twins! That's wonderful!' he would always jump in and say, 'Yes, they're IVF', as if it were a lesser victory. I didn't feel the need to broadcast the IVF element. These are our babies, regardless of how they were conceived.

I think IVF is fantastic. Years ago, we would not have had any other choice but to use donor sperm. How much things have changed. Hopefully, in another ten or fifteen years, with new technology, there will be more success for women over 40 and better screening for chromosomal abnormalities.

When the twins were about two years old, I decided to go back to the clinic and donate my eggs. As with our decision to donate our embryos, I wanted to share our joy. I couldn't believe how much Georgia and Hamish had turned my life around and how wonderful motherhood was. It seemed that every time I opened a parenting magazine I saw ads pleading for egg donors. It wasn't too much to ask. A few weeks of discomfort to change someone's life for the better is not a big deal. The woman I donated to wanted to know all the sorts of things about me that I'd wondered about our potential sperm-donors. I happily obliged and donated twelve eggs. They froze some, but the best were transferred fresh to the woman whose cycle was brought into sync with mine. Sadly, I don't think she had any success, otherwise I'd have been notified.

Around the same time, Colin had a nasty car accident and sustained a head injury. He became quite a different person. He lost his sense of humour and became very angry. He could not cope with the noisy demands of two-year-old twins. Normal toddler behaviour that had previously made him roll his eyes and smile indulgently now made him kick a teddy bear across the lounge room in a fury. It was no longer a healthy environment for the children. The crunch came the night I heard Hamish and Georgia screaming in the bath. They usually loved bath-time, so

I raced to investigate — Colin had them in freezing cold water. I thanked God it wasn't boiling water, but I knew then it was over. After separating, Colin coped much better with frequent, short visits from the kids. It was very sad, but we're still mates and he's very involved with the twins. He now lives on a houseboat on an invalid pension.

My fear of coping on my own made leaving hard. I worried about how I'd keep a roof over our heads, and keep the ponies and the lifestyle we'd enjoyed. As it turned out, the anticipation of leaving was worse than the reality. The day-to-day business of keeping everyone clean, fed, and well mannered was easier without Col in many ways. It had been draining trying to keep him happy, and avoiding dramas. On my own I had more control over our lives. It was busy, but I really enjoyed the children's company and was able to devote my full attention to them, and we became tightly bonded. I also had my frail, elderly dad living with us, and he was a great emotional support. He enjoyed knowing he was needed, too.

After a couple of years, and once the children were visiting Col regularly, I began to seek out adult company again. I started internet dating just so I'd have someone to go to a movie or dinner with. The last thing I was looking for was a new husband. For a few months I had a fabulous time, meeting new people and finding my feet again.

I nearly didn't contact Rob. His internet name was 'Mr Possum', which sounded a bit weird, but there were so many attention-grabbers in his profile I couldn't resist. He was only a couple of years older than me and we sounded so alike. It turned out 'Mr Possum' was a family joke, being the cat's name and also his video store password. The Vietnamese video shopowner thought that was his real name and always called him 'Mr Possum' — an unlikely name for a strapping, six-foot-three rugby player!

We exchanged a few emails, but it was our first phone

conversation that made a huge impression. I was in the middle of moving house, and Rob offered to help. I couldn't believe it. People I'd known for ten years hadn't offered to help, and personally I'd rather chew off my arm than move house if I didn't have to, let alone for someone I barely knew. Within two weeks, Rob and I were living together. I hadn't introduced anybody to the children until then, but felt very comfortable with Rob. They adored him, too. He is such a big kid. He likes to rough and tumble with them, and crams his long legs into their cubby house to share sherbets. He doesn't bat an eyelid at sitting through a four-hour ballet concert for Georgia's three-minute stage appearance, and he's always there for Hamish's soccer matches. He's brought such love and stability into our lives. Aside from the beautiful home and financial security, it is so much easier having two parents who relish the noise and craziness of children at the centre of our busy lives.

We were together about eighteen months when the baby bug slowly crept up and whacked us both on the head. Unbelievably, Rob had also had a vasectomy. I thought, *What are the odds of that?* The second thought was relief and joy that he wanted to try for another baby, too. Although Rob's only 41, he has three children aged twenty, eighteen, and fourteen. Every one of them was a happy accident. We have five fabulous children between us, and much to be thankful for already. Another child of our own would be the icing on the cake.

Rather than try to reverse his vasectomy, we decided to do IVF — partly because the embryo I donated had Down's syndrome, which gave me a bit of a scare. With IVF you can screen the embryos before they are implanted for Down's and some other disabilities. The other reason was that after a vasectomy reversal you often have to wait twelve months to see if it's worked. Sometimes the body produces antibodies to its own sperm after a vasectomy. Also, by this stage I was almost 39, so I didn't have time to muck around.

I went back to Monash IVF in Brisbane because it's such a family atmosphere and I knew the staff well. The odds of falling pregnant again so easily were much slimmer because I was six years older. On the other hand, I felt more relaxed this time around because I knew what to expect.

However, we weren't as well prepared, financially or physically. We were stunned at the cost. We hadn't been in our health fund long enough for it to cover much. After just one go we were $10,000 out of pocket. Rob also wasn't prepared for the pain of extracting sperm from his testicles. He has high blood pressure, and insisted on a local rather than a general anaesthetic. The surgeon wasn't able to be as thorough because of the discomfort, and we ended up with very little sperm.

Being older meant I had to take a heavier dose of hormones to stimulate my ovaries, too. Again I suffered homicidal mood swings and felt more miserable than during my first attempt. I also hyper-stimulated, which can be dangerous. I was bloated and had strong, crampy pain. I was on the verge of being hospitalised when I found enormous relief in drinking gallons of Gatorade. Somehow, that eased my symptoms.

Although I yielded thirteen eggs, only five fertilised. We wanted to screen the blastocysts for Down's syndrome, but the scientists weren't keen. One rang me from the lab and said, 'Are you sure you want to do this? You only have five embryos and the process can knock the cells around and sometimes stop them multiplying.'

It was a hard decision to make, but I insisted we go ahead and thank goodness we did — because one embryo turned out to have Down's. By the time we got to the transfer, five days after fertilisation, all but one of the blastocysts had fizzled out. That one was transferred into me and, because of the screening, I knew it was a potential baby boy.

Unlike when I fell pregnant with the twins, I didn't have any of the classic signs such as sore breasts or an increased sense of smell. I wasn't surprised when I started bleeding, but I was completely devastated. I cried for about a month. Rob was more stoic. He took the kids surfing and kept them occupied while I retreated into my shell and grieved. It really knocked me off my wheels, and I can't imagine how some women keep going back cycle after cycle. I didn't reckon with the emotional cost of failure.

If I had fallen pregnant, the baby would have been due around now. Rob bought a Dalmatian pup to distract me and fill the gap for a while. He's also been scouring the classifieds for a genuine mini pig, because I've always wanted one.

I know I shouldn't take these things too seriously, but a clairvoyant told me this IVF round wouldn't work but to keep trying because we would eventually have a little girl. We will go back next year and try again. By then, the medical fund should kick in and help cover some of the cost. We're also going to take vitamin supplements, get fitter, and Rob will switch to wearing boxer shorts to improve his swimmers. He'll also have a general anaesthetic next time and, hopefully, we'll get more sperm.

The twins are six now and are aware that some people need help from doctors to make babies. We have framed photos of them at five and seven cells on the mantelpiece, along with their first birthday snaps. They also know we're trying to have another baby, and Georgia has put in an order for triplets! In the meantime, we are enjoying the five children we have and are grateful that the IVF option exists.

CHAPTER 10

From Serbia with Love
–Anita's story

Anita's husband, Peter, was at first very resistant to trying IVF. Taking a break from their quest to conceive, the couple took a holiday to their parents' Serbian homeland and a monastery renowned for its miracles. But it was only when her husband got off the tour bus that he found some answers and got back on board the journey to parenthood.

My husband, Peter, and I met in high school. I was twelve and he was fourteen. He used to say, 'When I'm older, I'm going to marry you.' Both our parents are originally from Serbia. They're also pretty strict, so we didn't go out together until I was almost nineteen. We were married when I was 22 and he was twenty-four. We've been married eight years now.

We were very blasé about contraception. We didn't really focus on it, but about four years into our marriage I began to wonder if something was wrong. I went to the doctor and he said, 'Keep on trying. You should be okay.' So I changed gynaecologists a couple of times because I knew there was something wrong and yet no one was really paying me any attention.

My husband was also very relaxed about it, and back then he didn't talk about infertility. At one stage we were doing the

Billings method, where I would chart temperatures and look at my cervical mucus at the same time.

I found a female doctor who did a laparoscopy to check for endometriosis, but found nothing abnormal. Again I got the 'just keep trying' chat. So we tried for another awful year. I diligently charted my temperature and mucus for 12 consecutive months. I'm not a short-term person; I've always given every method a really good shot.

After more frustration I went to IVF Australia at Westmead, and they put me on Clomid to ramp up my ovulation. But I clotted really badly and had to stop after three or four menstrual cycles.

I was then referred to another doctor at the same clinic. I tried to tell her I thought something was wrong with my menstrual cycle. 'I spot every single cycle, and my period lasts for at least seven days, and intercourse is sometimes painful,' I said. 'There has to be something wrong; that's not normal.'

'No, you've had a laparoscopy and they didn't find anything, so there's nothing there,' she said.

'I understand they didn't find anything,' I said. 'But I just don't feel right. I don't wish to pursue IVF or IUI without a full investigation first. What's the point of doing IVF if my uterus is covered in endometriosis or cysts?'

The doctor wasn't convinced, but I persisted, 'For goodness sake, I'm paying for it. I want another test!'

'We'll book you in, and I'll look at your last test results, but if the report says there's nothing there, we're not doing it again,' she said.

Anyway, sure enough she rang the next day and said, 'I cancelled your booking. There's nothing wrong with you. We'll book you in for an IUI.'

I agreed, but I still had a feeling something was not quite right. My husband did a sperm test, which showed his volume was a little low and some were damaged, but it wasn't a huge problem.

I was sure the clinic had missed something, and I didn't like the way they treated me. My doctor didn't consult me at any stage, and I felt like a number rather than an individual. In any case the IUI didn't work, so I decided to try another doctor and clinic.

Some of the women on the Essential Baby on-line chat room suggested I see a female doctor at Sydney IVF for a hysteroscopy and laparoscopy, which I did. She cut into the lining of the uterus where she found severe endometriosis, which would not have been visible in a normal procedure and is why the other doctors hadn't found it. She even took photos of it to show me. Endometriosis is a disease of the endometrium lining, which means the embryo can't implant in the uterus and you can't fall pregnant. Rather than shedding normally every month during your period, the lining embeds itself elsewhere and spreads. I had two fingers' worth in thickness, which they had to cut out and drain. It was excruciating. When I woke up from the operation I hadn't eaten for 24 hours, so I was really dizzy. When I got up to go to the toilet I fainted on the ground with a catheter sticking out of me. I sat on the toilet and said to the nurse, 'If you don't give me a sandwich I am going to faint again.' I ended up sitting on the toilet eating a sandwich with my gown falling off. My dignity went out the window that day!

But overall the operation was a success, and I felt completely vindicated. My new doctor said, 'Wait three or four months. You may very well fall pregnant naturally after this, because we've got rid of all the endometriosis.'

So we waited and tried naturally, but nothing happened. I knew our next option would have to be IVF. But my husband wasn't so keen. I think he was fearful of something he didn't understand, and didn't even want to talk about it. He thought we should conceive 'naturally' or not at all. I thought it was best not to push it, so instead we decided to go to Europe for a holiday. We both needed a break. We went to Serbia and booked a bus tour to a monastery in Bosnia Herzegovina where people go to pray for miracles. As we were boarding the bus, for some reason my husband said, 'I need to get off. I can't make this trip.' I told him to go and I would call him as soon as I arrived.

At the monastery I saw the darker side of life. It was crowded with disabled children, blind people, and homeless wanderers, all looking for answers. It made me realise how sheltered my upbringing in Australia had been. At first I felt uneasy, and I rang my husband and begged him to come because I didn't know whether I could cope on my own for three days, sleeping on the floor amongst all those strange people. But I stayed and meditated and prayed for three days, and actually felt better for it.

After my husband had hopped off the bus he asked a man for directions. After chatting for a while, the guy offered him a lift in his car. They enjoyed talking so much the man asked my husband if he wanted to go for a drink, and Peter said, 'Okay. Why not? Anita's gone, I may as well.' The guy's house was a huge mansion, with an underground pool with heated tiles so that when it snowed the flakes melted and ran into the water. Peter was very impressed with the whole set-up. Running around the gardens were the man's two-year-old twin daughter and son. The guy asked my husband, 'So, where's your wife?'

'She's gone off to the monastery,' Peter said.

'Really. What for?' The locals knew pilgrims usually went there with a particular problem.

'Well, we've been unable to have children,' Peter said.

'See this house, it's a mansion,' the Serb said. 'I would have given up all this just to have my children, they are the most important things in my life.' He continued to talk to my husband about the joys of fatherhood and how IVF had made it possible. By the end of the evening, my husband's attitude to IVF had completely changed. He rang me, really excited, and said he couldn't wait for me to come home so he could tell me the story. And then he said, 'As soon as we get back to Australia, let's try IVF!'

It was meant to be. If he'd come to the monastery with me and not met that man, he may never have crossed the line and accepted IVF so whole-heartedly.

Our first attempt at IVF was heartbreaking because our cycle was cancelled halfway through. Despite the follicle-stimulating hormones (FSH), my follicles failed to mature fully, and the lining of my uterus wasn't thick enough. We tried to treat the first round as a test case but, of course, we were bitterly disappointed.

The second time around, we completed the cycle. When we did the ultrasound, my lining was perfect, and the number and size of follicles was much better after doubling my hormone dose. But, despite sixteen mature follicles, only half contained eggs, and five of those were abnormal. We ended up with only one viable blastocyst which, sadly, didn't take. The clinic wanted me to rest for a month before trying again so they put me on the pill to even out my hormones. Unfortunately, the pill makes me feel nauseous and throw up. And, of course, with those symptoms, people always say, 'You must be pregnant!' I feel like beating my head against a brick wall and screaming, 'No, I am not!'

At the moment we're starting our third round.

I told my mother we were thinking about IVF, but she's not very supportive about the idea. She has a friend who has cervical cancer and blames it on having done IVF. Once Mum said that, I decided not to tell her any more about what we were doing.

On the other hand, my in-laws have been fantastic. My mother-in-law has given me my trigger shots. Both her daughters have a thyroid problem, and she's been giving them daily injections since they were toddlers.

It's been hard for my husband, too, because he's not really part of the process; he's just on the sidelines, watching in frustration and worried about me.

For me, having low blood-pressure, the operations are extra hard work. Each time I've had the egg pick-up, I've come out of the theatre feeling dizzy, and have had to stuff biscuits in my mouth quickly for a sugar fix. If I don't eat straight away, I faint.

The doctors say I'm definitely a poor responder to the drugs because at 30 I'm relatively young, and to produce only three good-quality eggs is unusual. The next cycle we'll probably try putting me on Clomid from day one to four and start FSH on day three.

I'd love to be a mum one day, but I don't think it'll happen. Even after only having two cycles, something deep down inside of me is preparing me for never being a mother of my own children. It's really sad because I'm at the age now where whenever I go to a party there are always several pregnant women. My sister-in-law is also pregnant, and she and my brother conceived on their first attempt. She simply said to him one night, 'We both love kids; let's try tonight to start a family.' And, bingo, it happened. I'm happy for them because, honestly, anybody who doesn't have to go through what we've been through is blessed.

I'd have to say to people contemplating IVF that it doesn't guarantee miracles, and it's a good idea to have other options up their sleeves.

We've tried alternative therapies as well. I've recently stopped acupuncture after a year of treatment, and I also saw a Chinese herbalist who specialises in infertility. My mother heard about a healing woman in Wollongong who supposedly takes the evil away from you. I saw her every Saturday for six weeks. I took my husband with me a couple of times, and he said, 'Anita, are you crazy?' And I said, 'Shh. Just sit there!'

The woman sits you in a little tin shed surrounded by statues of saints, and blesses you. She's an old lady and very sweet, but nothing came of it. I feel a bit foolish now but, honestly, she had people queuing out the door.

I don't think we'll take the IVF much further. I just can't take any more. My husband and I did talk about using donor sperm or eggs, but we've decided against it.

Before we went overseas last year, we read about two children in a Serbian newspaper who lost their parents in the war. The article really caught my eye. When we arrived in Serbia I told my cousins I wanted to visit these two orphans, and they organised it. The little girl and boy, aged seven and five, live with their very poor grandparents in a tiny, one-room house with just one stove and one cupboard. The plumbing doesn't work and there's water all over the floor. They have absolutely nothing. They rarely go out because they don't have a car, and the grandparents are too frail to catch buses. It broke my heart. Despite that, the little boy in particular is brilliant and top of his class. We really loved meeting them.

It was hard to leave, and my husband didn't want to go. When we came back home we didn't talk about it, because it was too painful.

I asked my aunty in Serbia if it would be possible to adopt them. I don't know how difficult it would be, but I've started to investigate adoption procedures between Australia and Serbia. Understandably, the grandparents don't want to give them up at this stage because they are the only family they have left after the war. But they're not well, and when they die the children will go to an orphanage.

My husband and I feel that they need our help, and we'd love to give it. We've got everything we need materially. We're building a big house, and we've got our cars and good jobs, and this would be a chance to give these children a better life. But like that guy said in Serbia, 'We'd give it all up to be parents.'

Postscript

Anita finally fell pregnant with twins on the third round of IVF. However, at thirteen weeks one of the embryos stopped growing, and an ultrasound revealed that its heartbeat had stopped. The other embryo continued to grow normally. In June 2006 Anita gave birth to a baby boy, Daniel. Anita and Peter feel they are blessed, albeit sleep deprived. They still don't know why they found conception so difficult, and have decided not to do IVF again.

Peter was at first very resistant to IVF and felt he should have conceived his child naturally rather than by 'artificial' means. But he now says, when Daniel gives him his little crooked smile, that he is so grateful for his son he doesn't mind how he was conceived.

CHAPTER 11

The Gift of Giving
–Cath's story

An impulsive decision to donate her eggs to a stranger turned out to be the best thing for Cath, her existing family, and her new one.

I remember when I was young and first heard about people who couldn't have children, I found it a strange and surprising concept. I always assumed I would have children.

I met my husband, Cam, when I was 21, and we had our son, Tom, a couple of years later. I had just finished uni after four years of hard work studying agricultural science. It never occurred to me to work in my career while I had kids. Consequently, my degree qualifications went out the window, and now I work in a completely different field. But I don't regret not having worked as a scientist. I'm not the same person I was back then.

My daughter, Casey, was born 20 months after Tom, and in between I had a miscarriage. I didn't know I had conceived her because I was still breastfeeding Tom, and after the miscarriage, I didn't think I was cycling again. It wasn't until I was four months pregnant and felt the baby kick that I realised.

I would have loved more children, but my pregnancy with Casey was quite traumatic and it frightened the heck out of all of us. If it weren't for modern medicine we wouldn't be here. I had a grade-four placenta praevia, which meant the placenta had grown over the cervix and blocked off the baby's exit. The real danger was that the placenta was attached to the part of the uterus that dramatically expands during pregnancy. There's a huge blood vessel which feeds it, and if that stretches and ruptures, you can haemorrhage to death within minutes.

At six months the doctors wanted to hospitalise me for the rest of my pregnancy, and I said, 'I've got an eighteen-month-old child to look after. I can't do that.'

'We'll put him into foster care,' they said.

'No, you will not!'

I was instructed to stay home, and wasn't allowed to go anywhere by myself or be more than twenty minutes from the hospital. If I started to bleed I had to get straight to the hospital. As it turned out, I did have some bleeding three weeks before Casey was due, and was rushed to hospital, but it stopped and I spent the next three weeks there before having a planned caesarean. The whole thing was so stressful it put me off having babies for a while. In the meantime, my husband had the snip, so that was the end of that!

Ideally, I would have liked four children because that's how many there were in my family. But Cam has only one sibling, so he was happy to stop at two kids.

Even now, I think maybe I should have another child, but my kids are in year eight and ten in high school, and to go back to babyhood now would be crazy. That's one of the reasons why donating my eggs has been so wonderful. I have contact with the beautiful baby I helped conceive, but at the end of the day I can hand her back to her parents.

I didn't know Sally and James when I agreed to donate my eggs to them. Our only connection was a mutual friend, Sarah. I was at a party at Sarah's one afternoon and Sally couldn't be there, but she sent an announcement, which Sarah read out to the group, asking if any one would consider being their egg donor — or if we knew anybody who would, to let them know. Immediately, I thought, *That's me. I could do that!* It was as casual as if someone had said, 'Someone left this jumper at the BBQ last week; whose is it?' and I'd said, 'Oh, that's mine.' It was amazing. I just knew it was my call. It was really weird.

I'd never even heard of egg donation and I knew nothing about it. It was just meant to be. The funny thing is, I don't see that mutual friend much anymore. The whole thing has been our destiny; it was as if I formed a temporary friendship with Sarah in order to meet Sally, so we might get together and bring this little baby into the world.

It was a bit tough for my husband to accept at first. Our marriage had been going through a rough patch at the time. All marriages have their ups and downs, and we had just gone through quite a serious down. For him, this was another huge concept to deal with.

But egg donation helped us in a strange way. In Western Australia there is a six-month gap between when you say you will donate and when you actually can. During that time you are required to undergo counselling with and without your partner. This gave Cam and me a chance to work through our marriage problems as well. Sally and James had to undergo counselling, too, and then all four of us went together. We needed to make sure everyone was on the same page. Cam turned out to be very supportive and, for me, it's been one of the best things I've done with my life.

However, the actual IVF process of injecting follicle-stimulating drugs and having my eggs extracted was not particularly comfortable. Whenever a doctor says, 'There's only a one-in-a-100,000 chance of this going wrong, I think, *Oh no, here we go* — because it is amazing how often I've been that one.

I hyper-stimulated from the drugs and produced almost 30 eggs at the first extraction. I got so tired and sick I could hardly function. The whole IVF process was a surprise to me. When Sally, Cam, and I first went to the hospital in Perth to find out about the process, the nurse said to me, 'Okay, pull your pants down.'

'No way! If I'm having heaps of needles over the next two weeks, I'm not pulling my pants down for a practice,' I said. 'Sally, you pull your pants down — it's your baby. Show us *your* arse!'

But my husband said, 'I'm not injecting Sally's arse, I'm injecting yours. Pull down your pants, Cath.'

Finally I conceded and, as I was pulling down my pants, I said to the nurse, 'The needle doesn't go in all the way, does it?'

'No, no, it only goes in a little bit,' she said, and then behind my back she looked at Cam and mouthed the words, 'All the way!'

That's when Cam plunged the needle into my backside. I almost hit the roof.

The whole thing blew me away. I really had no idea of what IVF entailed before I committed to it. But even if I had known how hideous it was going to be, I would have still gone ahead. We live in Margaret River, so I wasn't being monitored as closely as I might have been if I was in Perth. At one point during the hormone treatment, the clinicians looked at my blood test and said, 'Whoa, your results are off the top of the chart. We'll back off the dosage.' But it was too late; my ovaries were overstimulated, and the damage was done.

I also had problems during the egg-harvesting operation, because I don't respond well to anaesthetics. The plan was that I would go in that morning, have the operation, and be out by three o'clock. However, it turned into a nightmare because whenever I got up to leave I would throw up and pass out. I was dizzy, and had hot and cold sweats. I tried to leave three or four times before, finally, the medical staff said, 'Forget it. Stay in hospital overnight.'

After the operation I had to sleep upright for seven nights. I was so bloated with fluid I looked pregnant. The doctor said, 'You can't lie down, because if that stuff gets into your chest cavity, you're poached.' So I built a massive pile of doonas and pillows, and each night my husband would climb the mountain and plant a kiss on me and climb back down to bed. I felt like I was on a trans-Atlantic flight for a week. That part was probably the biggest bummer, because I hadn't expected it. I knew to expect some discomfort leading up to the operation, but in my mind that was the end-point. It was horrible but, once it was over, it was over. Sometimes you have to endure a bit of revolting stuff to get good gains.

From the 26-or-so eggs they extracted from me, about sixteen fertilised, and they transferred two at a time into Sally over the next seven or eight months. We stayed in daily contact throughout the process. I knew exactly when she was going in for the transfer, and I'd send her 'pregnant vibes'. I was on tenterhooks during the ten days between the egg transfer and her first pregnancy test. For Sally, the waiting must have been far worse. It's really tough for people like her who undergo lots of transfers in order to get that one 'bingo'. It's draining and traumatic. The amount of stress she has endured in order to conceive is mind-boggling. I really respect her for what she has gone through to become a mother.

Sally finally got pregnant on the seventh transfer. She was in a hardware shop when the hospital rang to tell her. She rang me straight away and I screamed. My husband walked in while I was on the phone and thought someone had died. When he realised Sally was calling to say she was pregnant, he was really happy, too. It was a huge moment for all of us.

The feeling of being on tenterhooks was magnified 100 times during her pregnancy. Every day I prayed with all my heart, 'Hold on, baby, hold on.'

Sally and I talked to each other a lot during her pregnancy. When she passed certain milestones, such as three months and then six months, I felt an immense sense of relief. I thought, *Okay, from this point on, babies are born all the time and survive; we're almost out of the woods.*

Then it got really exciting. I was in Perth around the time Sally was due, and she rang me to say she'd been to the doctor for a routine check and he'd ordered her to go into hospital immediately for an emergency caesarean. She said, 'I'm going in right now!' I raced down to the hospital and massaged her feet while she was getting ready to go in to the theatre. During the operation I sat with James' mother in the waiting room, and finally James came out of the theatre grinning and said, 'Come on then!'

'What? What is it?' we said.

But he wouldn't tell us. We went in and saw their little daughter, Daisy, just minutes after she was born. Seeing her for the first time was such a fantastic feeling, I can't describe it. It had been such a huge journey for us to get to that point. It was so emotional and wondrous. I know there are probably a lot of unnecessary caesareans today, but I am so grateful that surgery exists — because without it my daughter, Casey, and I would be dead, and we wouldn't have Daisy either.

Daisy is such an angel. I have never come across another baby like her; it's almost as though she understands she's a miracle and has been put on this earth to be a really special human. She's gorgeous and has an incredibly sunny personality. Everybody falls intensely in love with her. My husband and kids love her to bits, too.

Although she is Sally and James' child, Daisy's really special to us because of who she is and how she came to be here. My daughter, Casey, especially has a deep connection with her. They love each other to pieces.

I don't feel like Daisy's mum. Sally is her mum; there's no

question about that. She carried and breastfed Daisy, and is there every day with her. But it's something we have to keep talking about. It really is a huge concept to comprehend, especially for Sally, who deals with it daily.

Daisy is special to me, in a similar way to how my nieces and nephews are special to me. I have a great connection with children. Sometimes I would prefer to spend time with kids rather than adults. Children have a refreshing honesty I enjoy. When I describe how I feel about Daisy to elderly people, they say it's like the way a grandparent feels about their grandchild. I feel as though she's part of my extended family. It's as if our two families have joined, and we honour that by getting together often and celebrating milestones together. I'm also Daisy's godmother, which I'm really happy about too.

When I look at photos of myself as a baby, I see Daisy. She looks like me and her father, James, and his family. When James and I first met there was a funny moment when we looked at each other and laughed at the whole concept. But it's not as if James and I had a baby; it's just that our genes combined to give Sally and James their daughter.

Daisy's now two and a bit, and if I don't see her for a while I miss her so much. At this stage she is changing so quickly. She has awesome language skills and can construct amazing sentences and sing a million songs. She's been counting to twenty for a long time. It's really exciting watching her grow and learn. I love spending time with her, soaking up her beautiful energy.

Sally and I have talked about when we should tell Daisy about how she came to be here. Obviously, she can't understand the concept now. But we don't think it would be fair to suddenly spring the whole story on her one day without warning. We are looking at putting together a kid's book, in the Steiner education fashion. I'd like to have some little dolls made up to help tell her story. It won't have a lot of details, but it will be a symbolic tale designed just for her. Sally can do it with her, and so can I when we're together. That will lay the foundation for later on, so she won't know a time when she didn't have that story. It will be a natural and integrated part of her life. It could be a great tool for other families in the same boat, too.

It's my greatest wish to have a relationship with Daisy for her whole life, in exactly the same way I want continuing relationships with all my nieces and nephews.

Unfortunately, my parents haven't taken to the situation very well. I haven't been able to talk openly to them about it. Initially I wrote them a letter, because I thought they should know, but my mother feels very strongly that what I've done is like giving up my child for adoption. I can understand where she's coming from, to some extent, but I don't agree with her. Putting up a child for adoption comes so much later in the process, because by the time the baby is born you already have a relationship with it. There is no question that you develop a bond with your unborn child during pregnancy. By the time Daisy was born, she, Sally, and James were already a family unit, which is quite different from Daisy growing in my womb and then going to Sally's family.

My parent's reaction has been a big disappointment for me. At first my brothers were a bit funny about it, too, but now I can talk about Daisy to them and show them photos of her. I hope my mother eventually changes her view. What we've done is very special, and it's a great heartache that I can't share that joy with her. At the same time, I can't force my view on her. My mother's got her own challenges to overcome in life, and if this is too big to deal with, then that's okay. Hopefully, when Daisy is older she may want to meet and talk to her. It would be a pity if that option were closed off forever. These things take time. Mum is concerned about the impact on my life and family if the arrangement falls apart and ends in tears. Of course it could go wrong, but many situations have the potential to become sad and tragic. For example, marriage has so much opportunity for heartache as well as joy. I hope in a few years' time that Mum will see we are all still happy, together, and well balanced, and then she'll change her mind.

Fortunately, I have a great relationship with Sally's mother and brother, who live in America. We bonded when they were here, and I've visited them in the States. I get on with James' family, too. I went to school with his sister, who is a year older than me, and we catch up occasionally.

When Daisy was about a year old, we all thought it would be great for her to have a sibling. Sally had the remaining two frozen embryos transferred into her, but they didn't take. Without hesitation, I said, 'I'll do it again, and we'll see if we can get another one.' The second time the doctors backed off on the hormone dose, to avoid me hyper-stimulating again — but I still produced around 27 eggs, which is a huge amount.

Just before the egg pick-up, I warned the anaesthetist that I'd had a bad response to the general last time, and he said, 'Don't worry, I'm going to give you a completely different anaesthetic. You'll be fine.' Anyway, the same thing happened. When I tried to leave the hospital I passed out and woke up with vomit everywhere. I had to be re-admitted through the emergency department. This time, Sally and my daughter, Casey, were with me for support, which turned out to not be such a great idea. Sally was traumatised seeing me in pain. She also fretted because she thought Casey was upset, which she wasn't, because she knows what I'm like after medication. Sally felt guilty and kept saying, 'This is all my fault.' In fact, I was more comfortable with it than Sally, because I knew what to expect.

From my second batch of eggs, about nineteen fertilised but, sadly, none of those took, and Sally didn't get pregnant. It appears it was my destiny to help produce only one baby for Sally and James.

For me, it felt right to try twice but not a third time. When the last embryo was transferred and didn't take, I went through a lot of grieving. I went to see Sally shortly after the last one and, as I got out of my car, she came to meet me and we stood in the street holding each other and crying. We grieved for the babies that never came. That was really sad, but life's not all joy and happiness.

The only way Sally can get pregnant again is if somebody else donates to her, but I don't think she'll pursue that.

Egg donation is certainly not for everybody. I have no doubt it was the right thing for me in that particular situation. I wouldn't consider donating eggs again to another family. I have done that for Sally's family, and my energy is focused in that direction. If anyone is considering being a donor, they need to come from a position of abundance and love as opposed to scarcity. I don't covet Daisy and I'm not possessive of her, though I love being with her. Anybody who donates eggs needs that type of attitude. They have to do it sincerely from the heart; otherwise it's not going to work. I think Sally is a sensational mother, and we have very similar approaches and attitudes towards child-raising. We even agree on the things that annoy us about other people's child-rearing! I say to her all the time, 'You are doing a great job; you are a fantastic mother.' Every one of Daisy's natural skills and talents has been brought forward because Sally does such a good job nurturing her.

I would find it really hard to accept if I donated eggs to somebody who fed their child Coca-Cola from six months and let them watch ten hours of television a day. It's important that, if you donate, you are generous of heart and spirit, and open and honest in your communication. If anybody went into it thinking, that baby is really mine, they shouldn't do it. I feel a huge love for Daisy but, at the same time, I'm glad she is Sally and James' baby and that they love her as much as they do. Daisy is lucky because she has a lot of people who love her dearly and want what's best for her.

Donating my eggs is one of the best things I have done with my life. It's one of the most special experiences I've ever had. Obviously, having my own kids was a life-altering event, and helping Sally and James to have Daisy has also been a watershed. It's part of what defines who I now am.

CHAPTER 12

Test-Tube Grown-up
–Rebecca's story

Rebecca is one of the world's first 'test tube' babies. She was conceived at Bourn Hall in England in 1983 when IVF was still in its infancy. In 2006 she was appointed a governor for the infertility network ACCESS. Rebecca has been instrumental in the media and public-awareness campaign of the federal government's review of assisted reproductive technology.

I owe so much to my parents. What they went through was amazing, and if it wasn't for them and IVF, I wouldn't be here.

Dad went to the UK in the mid-seventies to work for the BBC, and that's when he met my mother. They had my sister naturally, but then my mother used an IUD as contraceptive and it scarred her fallopian tubes, making her infertile. When Mum and Dad found out they couldn't have any more children they were extremely upset, because they'd always wanted a big family.

My father heard about something called IVF. It was the early eighties and the world's first IVF baby, Louise Brown, had just been born in 1978. She was conceived at Bourn Hall in Cambridge in the UK. IVF was heralded as a great triumph, but at that stage had only a five-per-cent success rate. After doing some research my parents decided to give it a go, which was considered a very adventurous move. At the same time, they put their names down for adoption. On the day my mother found out she was pregnant with me, she received a call from the adoption agency telling her she could have a six-week-old baby boy. They went ahead and adopted my brother James, and our family very quickly increased from three people to five. Mum and Dad say it was the happiest day of their lives.

In those days, women trying IVF stayed for two weeks at Bourn Hall, a beautiful old mansion in the countryside around Cambridge. The staff believed if the women were well looked after and fed nutritious food in serene surroundings, it would help their chances of falling pregnant. Mum felt very nurtured and cared for. One night she was sitting up late, knitting, and the nurse came by and said, 'Be careful. Try and rest. Think of the baby.'

In contrast, IVF clinics today are much more streamlined, and women only go into the clinics for their blood tests and procedures. It would be nice if women could still experience that luxury of being spoiled and looked after, because from what I understand IVF can be hard-going, quite clinical, and scary.

My Dad said Bourn Hall was very international in those days, because it was one of the few places in the world doing IVF. People would fly in from all over to get on their program. While Mum was having her eggs harvested, Dad sat in the pub down the road with all the other nervous husbands from as far afield as India and Africa, waiting for the clinic to call them to say it was time for their sperm.

Understandably, Mum was a little apprehensive about the IVF procedure. It was so new and no one knew much about it, and there was only a small chance of it working. Fortunately, my parents are very relaxed and don't dwell on the negatives. Rather than thinking, *Oh dear, I hope nothing goes wrong*, they focused on the potential and said, 'Wow, aren't we lucky to be part of this amazing new way to create children?' They were excited by the prospect of having another child.

Mum was 32 when she did IVF, and fell pregnant on her first attempt. I think her chances of success were higher than average because she'd already conceived my sister naturally.

I've got an old typed letter at home written to my parents from Dr Steptoe at Bourn Hall saying, 'Dear Mr. and Mrs. Featherstone, we've now conceived 100 babies at Bourn Hall and so far none have any abnormalities'. Obviously people in those days were concerned about IVF babies having birth defects.

Two years ago, Mum and I visited Bourn Hall, which is like a five-star hotel in Cambridgeshire. I carry a little picture of it in my wallet as a memento.

It was wonderful for both of us, because I'd heard so much about the place and it was a chance for Mum to reminisce.

When I got to England I rang the clinic and told them my history, and they said, 'What a shame. If you'd come a week ago you could have met Doctor Edwards, who helped conceive the first IVF baby, as well as you.'

The staff members were warm and welcoming. They gave us a cup of tea and showed us around. I even got to meet a newborn IVF baby. One of the nurses pointed out a small room and said, 'That's where your dad would have gone and looked at magazines while making his contribution!'

While I was there I tried to find out what number IVF baby I am ranked in the world. I was really disappointed that they couldn't find the old microfiche files. From the dates, though, I figure I was somewhere in the first 50 to one hundred. After Louise Brown, who was born in 1978, the next IVF baby was a boy, and then Candice Reed was Australia's first test-tube baby and the world's third. At that time, Australia and England were the only countries having any real success.

Every year Bourn Hall celebrates Louise Brown's birthday

with a party for all the people conceived there. I really wanted to meet Louise, but it was the wrong time of year.

I was born in 1983 when the general public knew very little about IVF. When my family came to Australia in 1985, Mum wanted to set up a support network for other IVF parents because there was nothing around then. But she had three young kids and didn't have the time.

Even today I come across so much ignorance. I recently did a live radio interview and the presenter said, 'I suppose because your mother couldn't have children, that means you can't either?'

'No, that's not the case at all,' I said. 'Louise Brown has a baby. And her sister, who was also IVF, was the first IVF child to have a baby naturally.'

I was twelve when Mum told me I was conceived through IVF. I remember we were in her room just hanging out when she first told me. I started to cry because I didn't fully understand. 'Does that mean I'm fake?' I said. Then she explained a little more and told me I was a miracle baby, and made me feel special. But I didn't feel comfortable telling any of my friends until my second year in high school. When I did, my close friends were excited for me and said, 'Oh wow, that's great.' But some other kids said, 'Really, are you sure? Are you making it up?'

Another girl said, 'I can't believe that. Don't you feel embarrassed? That's so wrong!'

Sometimes kids teased me and chanted, 'Test-tube baby, test-tube baby.' That hurt me. I feel as normal as everyone else; my parents just needed a bit of help.

Most recently, a 30-year-old woman, who should have known better, said, 'You're a freak!' when I told her I was conceived through IVF. That was pretty shocking. When people make those sorts of comments I usually say, 'I'm proud I'm a miracle child, and if it weren't for IVF I wouldn't be here'. That usually shuts them up. The people who say those things are just ignorant, but I'm still alarmed when relatively smart adults don't know anything about IVF. If it weren't for assisted reproduction so many couples would be childless. It's a wonderful technology, in my view.

I've now met several IVF adults in Australia, or IVFlings, as we call each other, including Australia's first, Candice Reed. We met through ACCESS, an infertility support network, and had a conference at St George Hospital in Sydney. It was fun because we swapped stories about growing up. We do feel a bit special, because our parents tried so hard to have us and underwent some scary procedures. Like me, a lot of them experienced teasing as kids. One guy said he didn't tell anyone until he was eighteen. I also met a girl who was Australia's first surrogate IVF baby, and she had a bad experience with one of her teachers who disagreed with surrogacy and ridiculed her in front of the class. It was great to talk about those things and share our feelings.

We all agreed it would be good if the concept of IVF were taught in school. Teachers could explain in layman's terms how the procedure works and why some couples do it. Being an IVF baby doesn't mean you're shoved in a test tube, put on a Bunsen burner, and mixed with sugar. People should understand that IVF is just another alternative to having children naturally or adopting. It's a positive thing, it's normal, and there are plenty of IVF children out there. If there was greater awareness it would become more acceptable, and kids wouldn't say hurtful things like, 'You're a test-tube baby, you're weird!'

I am surprised the pioneers, Robert Edwards and Patrick Steptoe, aren't household names. After all, they created the first IVF baby, and that's an important part of history.

I would like to have children one day. I know it's important not to leave it too late. There's been a lot in the media, and the message is getting out to girls my age that the older you are, the harder it can be to conceive. At this stage, though, I'm still career driven. I work as a promotions manager for a hotel chain and really enjoy what I do. It's a great company and there's a lot to learn, but I also want children. I think I could handle both.

IVF is a great option for childless people, but I wouldn't want to see it used for choosing height and eye colour — that's way over the top. If you want a child that much, surely it doesn't matter how tall or short he or she is.

I read in a magazine the other day about a 67-year-old woman who had an IVF baby. Now that, to me, is really getting out of hand. I think it's got to stop somewhere. Sixty-seven is just too old. You've got to think about the rights of the child.

I don't have a problem with sperm or egg donation, or surrogacy for gay couples. When you're desperate to have a child you will go to any measures. But I do think it's important the child has the option to get to know their egg or sperm donor.

This might sound silly, but I think I'm so determined and always try hard to get what I want because that's what it was like when I was conceived in that petri dish. There was one sperm and one egg that were stronger than all the others. In the race to fertilise the egg, I was part of the sperm that said, 'Get out of my way — I'm getting it!' It's funny, but that's the kind of person I am.

CHAPTER 13

Father Pride
–David's and Jason's stories

David and Jason have been in a committed relationship for 24 years ... but it wasn't until recently that Jason acted on his life-long yearning for a child. By taking advantage of the commercial approach to egg donation and surrogacy in the US, he bypassed Australia's restrictions on gay couples' access to IVF. They now have a baby girl called Gemma.

Jason: I've always wanted children. I think it's innate. When I was younger I thought, *Oh, well, I'm gay. That's it. I can't have children.* But later my paternal or maternal instincts took hold, and I felt I had something to offer as a parent. As I grew older I thought, *All this knowledge I have will be wasted if I don't pass it on.* Wouldn't it be wonderful to give a child not only material things, but knowledge and the chance to have a good life? I also wanted that wonderful feeling of being close to someone and giving unconditional love.

David: When Jason raised the subject of having a child, my immediate reaction was, 'No way!' It wasn't that I didn't want a child but, growing up as a gay man, it was just out of the question. It wasn't on the radar at all, so I put parenting right out of my mind and got on with life. That was just one of the things you had to accept as a gay man.

Jason: I remember seeing a TV news item about a gay couple from the UK who went to America and had twins through a surrogate. It was a big deal back then, and people debated whether it was legal or even moral. When I first heard about it, I thought, *I could do that, too,* and I started exploring the possibility. I also considered fathering a child with lesbians or through adoption. But, being gay, adoption wasn't available to me, and having a child with lesbians means you can't control your access to the child. We've heard horror stories about gays fathering children with lesbian couples and they have a falling out, then the women move to Darwin and the dad has no contact.

The only real option was surrogacy. I researched it on the internet and found an agency in Los Angeles offering a service to gay couples. So I flew to LA to check it out and make sure it wasn't some fly-by-night operation. I went to their office, and they showed me videos and records of their work. I even took a bank draft with the deposit. But I realised I hadn't really discussed it fully with David, so I told them I'd go back home and think about it.

In the meantime we saw a program on SBS TV called 'Two Men and a Baby' featuring the same LA agency. That confirmed to me they were legitimate. Prior to that I'd thought, *Oh, well, if I lose the deposit, I lose it. That's a risk I have to take.* After seeing that program, I said, 'That's it, I'm doing it.'

David: My greatest concern has always been my age. I'm 66, much older than Jason, and I thought, *When the child is going through adolescence, I'll be in my late seventies.* My number-one worry was, am I going to be able to deal with this and, number two, am I going to be a competent presence in the child's life because I'll be an old man? We didn't know if it was going to be a girl or a boy, and I kept thinking, *At 75, will I be able to kick a footy in the park with it, if it's a boy?* But I had to get my head around it, simply because Jason was going to go ahead and have a baby anyway. I either had to go on the journey with him or get off the train. When I knew how serious he was and how important it was for him, I was happy to say yes. But I had to set some boundaries by saying, 'I hope you're not expecting me to look after the baby all the time just because I'm retired and at home.'

Fortunately, Jason has been very good and has placed no expectations on me whatsoever. At the same time, I pull my weight; I do night feeds during the week because I want to help. But the agreement we have is there's no pressure on me; and when I knew that, I was happy to participate. But it did take a while for me to come around. I'm a Virgo, so I worry about the minutiae. But it's been fine. We've been totally on the same page ever since last December when insemination took place. And it's been a wonderful journey.

Jason: When I told my colleagues at work what we were doing some said, 'Why on earth do you want to do this? You have a good lifestyle. You're very comfortable. Why do you want all the problems that come with a child?' I said, 'I can't explain it. I know there'll be sleepless nights and other problems, but it's something I really want to do.' They've been very accepting, and actually gave me a baby shower!

It's been an interesting journey bringing little Gemma into the world. Firstly, the agency found a surrogate mother for me and then I had to choose an egg donor. The two are separate to make it easier for the surrogate mother to give up the baby. Eventually, they found Jane, from Ohio, who was 25, and a single mother with one child. The agency never chooses women who don't have their own children. Would-be surrogates are invited to an information conference where doctors and other medical experts explain the process and interview them to make sure they are psychologically and physically fit.

We wanted to be respectful of Jane. Some people try to tell their surrogates how to live their lives while they're pregnant, but I think that's foolish. How can you control someone miles away in a different country anyway? I wanted Jane to be on my side from the start. I told her I didn't want to interfere too much with her life and that it was wonderful she was carrying my child. I never made any demands on her other than she didn't smoke or drink during the pregnancy. She said, 'Don't worry. We can't smoke or drink alcohol, otherwise the agency wouldn't have chosen us.'

I rang her after each doctor's appointment, just to see how she was. I never asked, 'Have you been eating well? What have you been doing? Have you been exercising?'

David: There was no worry about her exercising. Jane's a postie, and walked eight kilometres a day until the week before the baby was born. We've heard of people making terrible demands on their surrogates and ringing them every week to interrogate them. Jane says she'd be a surrogate for us again, but not for anyone else. She's spoken to other surrogates who've had lots of demands put on them, and realised how lucky she was.

Jason: When it came to choosing the egg donor there were three different agencies to choose from, with hundreds of profiles and photographs of women on their websites.

David: Jason would be upstairs on the computer going through the profiles and he'd say, 'David! Come and look at this one!' I'd come up and he'd say, 'What do you think of her?' and I'd read the profile and say, 'Oh, she's all right,' but I still wasn't 100-percent committed to the idea back then and was hoping the idea would fizzle out.

Jason: When I realised David wasn't interested, I went ahead on my own. It was a bit lonely at first, because I found choosing the donor the most difficult part. I didn't want to make a mistake. The surrogate is not a problem, as long as she's healthy. But I wanted a donor who was intelligent, and preferably someone studying science, because that's my background as a doctor. Most of the donors are students or out-of-work actors, and have to be under 27 years of age. The first one I chose was studying at Brown University. But by the time I wrote to the agency, she'd already been taken. I found it very difficult, because there were hundreds to choose from. They give you a profile of all their physical attributes, family and medical history, IQ and university entrance scores, and the course they're studying. Some agencies even have videos of the women being interviewed so you can see their mannerisms. Most of them don't have children, because they're so young. Finally I chose a psychology student. Like most of the donors, she decided to donate her eggs to help pay for college fees.

Our donor produced 25 eggs and we got fifteen viable embryos. I went to the US to deposit my sperm. You can FedEx it if you want, but it has to be frozen, and costs around $2,000 to transfer to Los Angeles. I'd rather they used my sperm fresh, so I went over there for a day. The sperm is then tested for HIV, Hepatitis B and C, and other diseases.

The lab updated me daily by email, telling me how many eggs had been extracted and how many had fertilised.

Two embryos were transferred into Jane, who travelled from Ohio to LA and stayed in a hotel for a week for the procedure. I thought she might be bored in her hotel room, so I rang to chat and make sure she was okay. She said, 'Well, I think things are fine, but you can't tell at this stage.'

After her first examination and blood test she sent me an email to say, 'Yippee! I'm pregnant!' The lab also sent me the results of her blood test, and because the hormone level was very high I knew it meant twins. It was one of the happiest moments of my life. It was like when I got into medicine. You always remember those moments. But I was cautiously optimistic because I know things can go wrong, so I didn't break the news to David immediately. Then I got another email from the lab telling me Jane's hormone level was continuing to rise and I said to David, 'Prepare yourself. We may have twins!'

David: Aarrgh!

Jason: But a week later the lab sent me the latest hormone results and they had decreased to a normal level, so I knew one embryo had failed. Jane sent me the photograph of the ultrasound and said, 'You can see there's a blank' — she called it a 'blank one' — 'and there's a full one'. That's the way she described the embryo and the blighted ovum.

We planned to go to the US for the eighteen-week ultrasound which would determine the sex of the child. But in the meantime Jane sent us the photographs and told us it was a girl. We decided to go over anyway and meet her face-to-face for the first time.

David: We took her out to dinner, and she brought her mother along for moral support.

Jason: Her mother said, 'Jane didn't know what to expect of two men, from so far away.' Then after dinner she confessed, 'Now I've met you both, I feel so much better. I was worried you might turn out to be really horrible and my daughter would have to carry a child for two awful men.' She also admitted she hadn't wanted her daughter to do surrogacy in the first place.

David: But Jane's got a mind of her own. She was determined to do it, and wants to set aside the money for her son's education because she's doing it tough as a single mother.

Jason: What makes me crazy, though, is that the surrogate doesn't get as much money as the agency, so we gave Jane some extra cash because we feel it's a wonderful thing she's done and a huge sacrifice.

I don't think she really enjoyed her pregnancy. She would have preferred to be the egg donor. But the agency convinced her to be the surrogate because by the time she would have gone through the screening process, put her photographs on a website, been selected by someone, she would have been over 27. So they said, 'There's no point. Why don't you be the surrogate instead?'

During the pregnancy I couldn't relax, because I was frightened of possible abnormalities. I kept thinking, *What if something goes wrong?* I asked the obstetrician to do the antenatal screening for Down's syndrome and other congenital defects, but he wouldn't. I knew Jane and the egg donor were young, but I would have felt much better if they'd done the trisomy test for chromosomal abnormalities. The obstetrician said, 'In America, we go according to the insurance. If the insurance company doesn't want to pay for it, we won't do it.' I said, 'But I'll pay for it.' He said, 'Oh no, you can't do that. It's illegal.' I still don't understand how their system works.

Even at the birth I was nervous. Jane was induced and had a normal vaginal delivery after a ten-hour labour.

David: I wasn't there. I would have liked to be, but she already had one stranger there and I didn't think it was right to have two strangers looking at her while she was giving birth. Her mum and Jason were with her, and I thought that was quite enough. So I just stayed in the motel.

Jason: I have delivered quite a few babies myself, but because it was my own I was so worried. I kept thinking, *I should be happy, but I'm a nervous wreck*. When Jane was pushing, the obstetrician said, 'I can see a lot of black hair!' And I said, 'Well, that's my child then!' As soon as Gemma was born, she cried, but only for a few seconds. I thought, *Why has she stopped crying?* I checked her colour and examined her legs, hands, and everything. I was still wearing my doctor's hat. But as soon as they gave her to me and I knew she was fine, I relaxed, and suddenly a warm feeling came over me and I thought, *Oh, isn't this the most wonderful thing!*

Jane had said from the start, 'I would like Jason to hold the baby first.' I held her for a few minutes and then said, 'You can hold her, too,' and I passed the baby to Jane. But she showed no attachment at all. She said casually, 'I'll have a look. Oh, yeah. She's fine. Here you go,' and she handed her back.

I had no concerns about Jane being attached to Gemma. I knew she had no legal rights over her, because that's all handled by the agency's lawyers and, being a single mother, she wouldn't want the burden of another child.

The one thing that was a bit upsetting, though, was that Jane wasn't prepared to breastfeed at all — not even to give Gemma some colostrum on the first day.

David: She didn't give us a reason and we didn't want to push her. Jason tried to explain how important colostrum is, but she still said no. She didn't even want to express any. Her mother may have had something to do with it. She couldn't get it through her head that this child wasn't part of her daughter. She felt she was losing her granddaughter. I wonder if she thought, *If Jane breastfeeds Gemma, that's a part of her going into the baby*, and it would make it harder for her to give her up.

Jason: At first, having a baby was quite overwhelming. But the nurses were good. It was a very small hospital in Dayton, and the staff were very nice to us.

David: We couldn't fault the hospital. But they were curious about us.

Jason: They were very curious! The social worker said, 'We've been waiting for you to come. We've been talking about you. This is the first time we've ever done anything like this at this hospital.'

David: The nurses took us for one session, with another couple of new parents, and showed us how to fold and change a nappy.

Jason: Fortunately, another nurse went out of her way to help us during her own time. Gemma was separated from the surrogate by then, and the nurse said, 'I'd like to teach you a few things.' She showed us how to feed and burp the baby. We learnt more from her than from anybody else.

David: The funny thing for me was that, all of a sudden, here was this baby. Not having the surrogate here, but rather miles away, we didn't experience the full journey of the pregnancy. It's not like having her in your own home and seeing her gradually growing belly. To us, it seemed like one day Jane was impregnated and the next, we've got a baby in our arms. It was a very odd feeling, because we didn't have that transition. I didn't feel connected to it all the way through because it was so far away. All of a sudden, boom! We've got this tiny thing on a motel bed crying and needing to be fed.

Jason: In the meantime we had to organise a passport for Gemma to bring her back to Australia. On her birth certificate the surrogate and I are listed as the parents. We don't know the donor's surname, so we couldn't put her down. Some people in our situation leave the mother's name blank or put two fathers. But I don't want Gemma having to constantly explain her situation as she's growing up. Every time she produces her birth certificate I don't want people asking, 'Who is your mother? Why have you got two male names on your birth certificate?'

A lot of gay dads in the States hire a lawyer through the agency to change the birth certificate so they can have two male names on it. I think the agency was surprised when we said we were happy to put the surrogate's name down as Gemma's mother.

Legally, Gemma will be able to contact the donor when she reaches eighteen. But I'd really like her to have contact earlier, maybe when she's at school. I want Gemma to know where she comes from, but to date we still haven't had a reply from the donor.

David: We can't talk to the donor directly; all communication is through the agency. We asked the agency if the donor would like us to send a picture of Gemma, but we haven't had a response to that either.

Jason: I don't think they're used to that sort of request, because most gay parents don't want their child to contact the donor. But I believe it's important for children to know their roots. They are always going to ask, 'Where did I come from? Who's my mother?'

Of course I'm concerned about the psychological impact on Gemma as she's growing up. I'm not so much worried about what other people think; as long as she's comfortable with herself, she'll be fine. But I think it's important for her to know where she comes from and who her mother or the donor is. I have my doubts sometimes, and worry she may grow up feeling displaced and say, 'I don't know who my mother is. Why don't I have a mother? I've never had a mother. My friends all have mothers. Why don't I?' When she's old enough I will tell her the truth: that she does have a mother, but we wanted her so much we took her mother's eggs and conceived her a different way. I want her to feel she is a wanted child, rather than a child growing up with two daddies and no mother. When I went to the agency, the first thing I asked was, 'When my child grows up, how do you think I should explain this to him or her? What do other couples do?'

'Just tell your child he or she is special and has two daddies,' the agency woman said. I found that hard to understand.

'No, I won't do that,' I said.

'Well, that's the only way you can do it,' she said.

Who says which is the best way? That is something I still struggle with and will have to deal with when the time comes.

David: When we were bringing Gemma home from America, quite a few people in the airport looked at her and said to me, 'And I suppose you're the grandfather?' If that was said to me once, it was said to me one hundred times. So that's when I said to Jason, 'I will simply be "Grandpa".' We'll deal with the rest later, when she is old enough to comprehend my relationship with Jason. We didn't want the child to have two dads. We think that's too much pressure on the kiddie. Although I am co-parenting with Jason, I'm not 'Dad'.

The way the law stands, I have no custodial rights to Gemma if anything happened to Jason. In that case, Jason's younger brother would become her guardian. He and his wife have an IVF child, too. I'm comfortable with that because I know that, even if we did separate, Jason is not the sort of person who would use the child against me.

Jason: I think, eventually, paid surrogacy will become legal in Australia. At present, Australians have to go to the States to do it.

David: I find it unfair that there are people who want to do what we did and would make fantastic parents, but are simply shut out because they can't afford to fly to America two or three times and pay all the expensive agency fees.

I think most people are very supportive of somebody who wants to have a child and has to go down this route. It's only a matter of the legislators changing the rules in Australia. I certainly don't think it's going to be an election issue; people aren't going to march in the streets with banners protesting, apart from some right-wing Christian groups. On the whole, I think people would be supportive. After all, if people want a child so badly and they're prepared to go to a lot of trouble, obviously that child is going to be very loved.

Jason: We told our gay friends and even some of my patients about Gemma and how she was conceived. However, I have certain friends in my hometown who I haven't told because they wouldn't understand. I've just said, 'I have a child. I'm a single father, and I don't want to elaborate on the details.'

David: Not everyone has agreed, of course. My sister has four children and is very Catholic. She said, 'I don't agree with it, but I will be supportive.' And she has been very good. I have two other brothers, who've been marvellous. No one has given us a hard time or said anything upsetting.

Jason: We've been very supported. Even our local maternal health nurse has been wonderful.

David: She's in her thirties with kids of her own, and is lovely. All the other girls at the centre with babies have been fine, too. We haven't experienced any prejudice, not even a look or a snigger. Maybe they do, behind our backs, but certainly I've felt very comfortable there.

Jason: When we first got Gemma, I took a month off work, which included our two weeks in America for the birth. Fortunately, I have a good support network, my mother looks after Gemma during the day, and David shares the night feeds.

David: We have a system. I do the night feeds during the week and then Jason does it on Friday, Saturday, and Sunday because he only works a half-day on Monday. It means he can go to work refreshed and not fall asleep in front of his patients. The broken sleep affects me more because I'm older. Once he goes to work and Gemma goes to her Grandma's, I have a nap and catch up on my sleep. It works well.

Jason: Being parents has obviously changed our lives. We certainly don't go out as much as we used to.

David: But I don't miss it, funnily enough. Gemma is the centre of our lives now. Once upon a time we had stimulating and intellectual conversations, and now the conversation revolves around, 'Did Gemma do a poo today?' I think, *Gosh, where has all the intelligent conversation gone?* That was terribly important in our relationship. But I guess it will change over time. We couldn't have done it if I was still working. I'd probably be thinking entirely differently about the whole thing. We're lucky I can stay at home.

When we first bought Gemma home, we were having dinner, and Jason leaned across the table and said to me, 'I must tell you how extraordinarily happy I am. I feel complete for the first time in my life.' I nearly burst into tears because I could see how happy he was. It was an incredible journey leading up to her birth; and once Gemma was born, Jason was transformed. When you see him holding Gemma and the love he has for her, it's incredible. He is the most devoted dad. It makes me so happy, seeing that. How could I have possibly denied him the chance of having a child, when I now see how much it's meant to him? Of course, I benefit, too. Gemma's gorgeous and it's lovely to have her here, but I can't possibly have the same feelings for her because she's not my child. Jason's feelings for Gemma are extraordinarily deep, and it's lovely to see.

Jason: I think this is a great stage in our lives to have a child. Some people say it's better to have a child when you're much younger; but, for us, we've done a lot of travelling and worked on our careers, so now we can settle down and have a child. We still have a dozen frozen embryos in the lab. Maybe we'll have another one transferred next year!

CHAPTER 14

Single-Minded
–Janine's and William's stories

Janine was 41 and single when she decided that, although 'Mr Right' hadn't shown up, she didn't want to miss out on having a child as well. A friend put her in touch with William — a gay man in a steady relationship who wanted to be a father. With the help of IVF, the two are on the bumpy path to pursuing their vision of parenthood.

Janine's story
I had an epiphany, almost two years ago now. While I was waiting for a girlfriend near the beach one day, I sat and watched three little girls having the time of their lives playing on the swings. I watched them for half an hour and thought, *Is this what I'm missing out on? Am I absolutely crazy? What am I going to do?* The idea of being a single mother grew in my mind, but I felt conflicted. I had several sessions with my kinesiologist, who reads emotions in the body, which is often a good way of bypassing the mixed messages we get from our brains. She confirmed that my body and soul really wanted to have a baby. But it took me three months to make the decision; it was such a big step. I looked at all the pros and cons, and cried a lot about not meeting Mr Right

and my shattered dreams of marriage with children. When I was growing up I assumed I'd be married at twenty-two.

I've had lots of boyfriends, but I've just never met the right person. Because my parents had such a bad marriage, I didn't want to marry just anyone, because you have to spend the rest of your life with them, waking up with them every morning, cooking with them, going out with them, and being at home with them. My parents' marriage was very angry and violent. There was a lot of yelling, screaming, and shouting, with little love or affection.

However, I have two brothers and a sister, all with children, and they have great marriages, with the usual ups and downs. Some of them have been married for more than 30 years. I, on the other hand, have never come close to meeting the right partner.

I realised that if I was to have a child on my own, one of the best options was IVF. I talked to my GP and she said, 'What you want to do is not uncommon. Single women do it all the time, particularly in their forties.'

I went to an IVF clinic with a sperm bank, and underwent a series of counselling sessions. The next stage was to pick a sperm donor from their list of profiles. The information given is very limited. I've always wanted a dark baby, perhaps because of my Greek background, and there was an Egyptian doctor on the donor list. I thought, *Wouldn't that be a beautiful combination?* But, unfortunately, they'd run out of his donated sperm.

My second choice was a Mediterranean guy, but he'd only donated enough sperm for five women, and his last sample had been given away the day before.

I spent the Easter weekend bawling my eyes out before coming to terms with it. Meanwhile my friend Kate went to a dinner party on Easter Sunday, and a gay friend of hers, William, was there, and said one of his regrets in his life was not fathering a child. Kate said, 'Well, you need to speak to my friend Janine.' When she told me about William, I thought, *The universe is looking after me. If this guy wants to be involved in bringing up a child, that's perfect, because then I won't have to do this solo.* I would much rather share the parenting than use an anonymous sperm donor. I also think it could be traumatic for a child, not knowing who its father is. Although the clinics keep all the donors' details on file, the child can't contact them until he or she turns eighteen. As a result, there are fewer men donating sperm, because they aren't guaranteed lifelong anonymity.

When I met William I really liked him. He's 51 and has been in a stable relationship for fourteen years. We talked a lot over three months. We went back and forth, debating and arguing the issues of parenting, and also hugging and laughing before finally coming to an agreement. William wanted to draw up a contract through his lawyer, which was absolutely fine with me. It covers details about contact with the child and how much he'll contribute financially. He wants to be hands-on, but this contract is for the worst-case scenario if we have an awful fight and aren't getting on.

After we'd thrashed out all the issues, I started trying to get pregnant by self-inseminating at home. I would go to his house in the morning before work and wait downstairs while he masturbated in his room upstairs. He would leave his sperm out for me in a cup and I would use a turkey-baster and insert it immediately.

For the first couple of months I used a little plastic syringe but was getting nowhere, so I bought a turkey-baster from a kitchenware shop because it's much longer. I also bought a kit from the chemist to chart when I was ovulating. We thought we'd try the self-insemination method for about six months, and if we had no luck we'd move on to IVF.

It's been really disappointing. There have been times when I was sure I was pregnant for about a week to ten days, but then I'd get my period.

The doctor said to me, 'Well, you can't have been pregnant, because the pregnancy test wasn't positive.' But I've had the funny taste in my mouth, massive boobs, waves of nausea, and all the classic signs of pregnancy.

William and I finally went to the IVF clinic and were told we had to freeze and quarantine his sperm for six months. By law, because he's my donor and not my partner, he has to undergo a series of blood tests to check for HIV and other diseases.

I was devastated. If I had known we'd have to wait six months, we would have started IVF much earlier. Now I'm 43, I feel I'm running out of time. During the quarantine period, we continued using the turkey-baster method, but still had no success.

At the clinic we initially opted for IUI (Intra-Uterine Insemination), where they insert the sperm directly into my cervix. The advantage of IUI over IVF is that it doesn't involve drugs; however, the success rate is somewhat lower.

Unfortunately, we had a series of false starts. We went to the clinic in January, but then William got sick and couldn't donate his sperm until the following month, and then it was August by the time the sperm was out of quarantine.

Finally, we were ready to go ahead. I'd had my counselling and was booked in for my blood tests to verify I was ovulating. Then, on the eve of the big day, the clinic rang to say a new law had just been passed, and William would have to undergo more hereditary disease tests — which meant we'd have to wait even longer. This time I saw red! I have never been so livid in my life. I ripped shreds off the nurse. If I'd been in the clinic all week having blood tests and there was suddenly some new law regarding sperm donors, why hadn't they told me about it earlier?

I said to the nurse, 'Okay, I can get William to have these tests tomorrow. But you've got to promise you'll turn the results around in one day.' And she said, 'I can't guarantee that.'

'Why not?'

'Because it's up to the pathology centre, and we can't tell them what to do.'

'This is an extreme circumstance. Surely you can mark it "urgent" and get them to turn it around in 24 hours.'

'I can't do that,' she said.

'Okay. I'm talking to the doctor.' So I called the doctor and ripped shreds off him, too.

'You've got to turn this test around in 24 hours,' I said.

'I can't. The tests take three weeks to come through,' he said.

I admit I was surprised at the intensity of my anger.

So William had to go in and have more blood tests. He was frustrated with the delay, too, but was more concerned about me because I was behaving like a madwoman.

Finally we did the IUI. At the same time I was so busy at work and under a lot of pressure, and began to panic again and ask myself, 'Can I really work and have a baby?' The old fears resurfaced. So, of course, I got my period again. We did IUI a second time and I really felt pregnant this time because my breasts were enlarged and tender but, no, I got my period yet again.

Consequently, we have decided to go the whole way and do IVF, which has a higher success ratio than IUI because they fertilise the eggs with the sperm outside the body before implanting the embryo.

It's been such an emotional roller-coaster. My intuition tells me I've been pregnant, albeit briefly. I've had so much fear surrounding the issues of being a single mother, having a child with a gay father, working and bringing up a baby on my own, that I believe every time I fell pregnant I'd panic and lose it. I also had worries about the child being made fun of at school for having a gay father. Kids can be cruel about anything. I used to stutter when I was at school, and was teased. I also worry about telling my mother, who's a strict Catholic.

But I think I've come to terms with all those fears now. I've done a lot of work to help me let go, and trust everything's going to be okay. I'm meditating to alleviate my anxiety, but still that little voice in my head wakes me up at three o'clock in the morning and gives me all the worst-case scenarios. Having a baby is a big thing, and I never thought I'd do it all by myself.

Fortunately, my sister and brother have been amazing, although my sister-in-law doesn't agree with what I'm doing. She thinks being a single mum will be really hard work. She has three children, so she knows what she's talking about.

William's partner, Tim, has also been very encouraging because he knows how much it means to William.

I'd like to take six months off work when the baby comes, and after that I'm just going to have to get a nanny. I'm happy for the child to go into childcare when he or she is two years old so they can play with other kids, but babies need to be looked after one-on-one. All my family live interstate, so unfortunately I don't have support.

The thing that keeps me going is when I look into the future and I see myself aged 47 or 48, I don't want to be childless. I want the experience of being a parent. I'm a very loving person. I'm going to be an amazing mother, and William will be a fantastic dad. I've been waiting for this next stage of my life for such a long time now.

I recently went to a picnic with some friends, and there were lots of kids running around. I spent most of the time holding a tiny baby, who fell asleep in my arms; it was gorgeous. My friends know I'm trying to get pregnant and they're extraordinarily supportive. I couldn't do it without them.

The other thing that gives me strength are the many single mums out there who have either chosen to do it by themselves, or their marriages haven't worked out and they've ended up on their own with kids.

It's now two years since I first sat in the park and watched those little kids playing on the swings. And I never, ever thought it would take this long.

Postscript
On my first round of IVF the doctor picked up five eggs: three were mature and only one lasted the distance. I keep telling myself, 'You only need one good egg!' When they implanted the embryo, I had an overwhelming feeling of happiness and euphoria. It was very beautiful and took me by surprise. I find out the result in a couple of weeks — fingers crossed!

Email
Just to let you know that my period arrived yesterday — I'm devastated. It has taken days for the enormity to sink in. It never occurred to me that I wouldn't be pregnant because I was certain IVF would work. I'm rethinking a lot of things and don't know whether I will continue with IVF. This whole journey has taken me by surprise. I don't look or feel my age, and I'm very fit and healthy, so it's a harsh reality that the age of my eggs is letting me down. My message for any woman in her late thirties who thinks she's got lots of time left is, 'Start now!' I'm praying and meditating and giving the problem to God, because the decision on whether or not to continue is too big to make on my own.

William's story
For a long time, as a gay man, I never felt I could be a father, and was quite happy to let that one slip by. In some ways I felt defined by not having a baby and the freedom that brings. I believed one of the important things about being a gay man was to use your freedom to do the things that the baby-makers couldn't do. So I've travelled a lot, had many different careers, been very promiscuous, and had a good life. It was only later, when I was in a secure relationship and at a particular age, that the idea of having a child struck me.

It was in the early nineties that a female friend and work

colleague first inspired me with the idea of fatherhood. I became very close to this woman, and she suggested I father her child using a turkey-baster. I fantasised about the meshing of our households and the involvement of my partner, Tim, and the happy families we'd play.

I broached the subject with Tim on our way to the coast one weekend. He was very supportive, and made lots of jokes about being my nanny. He's more domesticated than me, so there was probably a grain of truth in that idea.

When I told a gay male friend about my plan to father a child, he said, 'Oh, William. You think you can have everything.' And I said, 'Yes, I can have everything. I've had a very lucky life and this would make it absolutely perfect.'

But, for various reasons, the idea went to sleep. The woman decided not to go ahead with it because I think she wanted me as much as she wanted the baby. We're still good friends, but I think even now she mourns not being a mother. However, the idea of becoming a father continued to niggle in the back of my head.

Some years later, I was living in Canberra, where it is legal for people of the same sex to be the parents on their child's birth certificate. I was struck by the new gay and lesbian realities of this 21st century, where everyone was suddenly having children. I went to gay parties, and there were kids running around everywhere. That would have been astonishing 30 years ago. A lesbian couple who wanted a gay man to father their child approached me. They were a long-time couple, and I was attracted to the vivacious, artistic, younger woman who wanted the baby. We talked closely and it was a good conversation. We were going to talk further, and I left it to them to contact me and they never did. It was a very busy time for me, moving jobs and cities, and so I let that opportunity slip. In hindsight, it wouldn't have been an ideal scenario because it would have meant siring a child and letting it go to two devoted mothers. My input would have been minimal.

I now have more of an understanding of what it would mean to be a hands-on father, having discussed the subject with Janine. We had met briefly years ago at a weekend workshop in Byron Bay about warriors and goddesses. It was aimed at unleashing the masculine and feminine in our respective selves, which was quite interesting for me as a gay man — but that's another story.

Janine struck me as a very attractive princess, but it wasn't until a few years later that a mutual friend got us together when I mentioned my desire to father a child. We met and talked the idea through. I've been rather methodical about it, in a way I think that initially made her slightly fearful or chilly towards me. I asked her about her genes and background and why she hadn't been in a relationship. Was it because her parents as well as her grandparents had broken up, and there had been some violence in both families?

That made me a little nervous, as I'm sure some things about me concerned her, such as my family's eccentricity, which has its good side, but also its compulsive-obsessive side. We shared this stuff like two lovers on a first date. There was excitement and laughter, but with a degree of methodology as well. Tim had to take Janine aside after our first few meetings and explain to her, 'William's not cold; he actually feels this quite intensely.'

It was almost romantic. Janine and I had a series of lunches, and talked openly about our vision. We wrote down points to consider and a summary of the things we'd covered. We agreed the child would have my surname, and Janine would have the right to choose the first name.

We have a contract, which looks at financial roles, access, and the implications in case of a breakdown in our relationship. We've thought and talked it through very deeply and thoroughly. The contract covers all sorts of issues, well projected into the future.

I've talked about it to a lot of friends, and their reactions have been astonishingly varied and interesting, particularly from the women. One lesbian woman can't understand why my parenting role will be anything less than 50 per cent in terms of access, responsibility, and money. She doesn't understand that it's a negotiated kind of relationship. What's so magic about 50/50 anyway? My involvement as a father will not be absolute and 100 per cent, just as I'm not absolute and 100 per cent with Janine. I don't live with Janine, so it can't be that, anyway. My journey to gay identity has helped me resist the usual heterosexual expectancy of behaviours and the belief that these things are fixed in stone. We are forging something different here.

Janine and I would continue living separately. She may move closer to me for better access to childcare and education, and to make it easier for picking up the child from school while we still run our professional lives. We have agreed on certain financial arrangements over the major costs, including his or her schooling. The agreement also covers access. The lawyers got terribly pedantic about that and wanted us to stipulate the exact time on Christmas Day when the child has to go from one parent to the other. They even wanted us to nominate how many of my birthdays the child will spend with me. It's all a bit odd, but very defined the way lawyers like it to be.

I see the arrangement as liberating. It's as if Janine and I are starting with a divorce. The paper is there if we are in discord; but we're not in discord and we probably never will be, since we've never been in accord, in a loving, romantic way.

Tim has also attended some of these meetings, because obviously it's going to impact on his life greatly, too. One thing that is still unresolved is whether he will be present at the birth. It's agreed I will be present, and I suggested Tim might also be there, but Janine said, 'Oh, no, no. It wouldn't be appropriate. The room's too small.' I had to chastise her and say, 'I hope Tim didn't hear you.' Fortunately, he was in another part of the house at the time.

In the initial conversations with Janine, Tim was quite happy to be involved, and I sought his advice all the time with the negotiations.

When we were trying to get Janine pregnant with the turkey-baster method, I had to wake up at seven o'clock in the morning, masturbate, and leave my little offering in a glass. I always tried to choose a small glass that made my pathetic little emission look larger! I'd leave that by the bedroom table, and Janine would come in and give me two air-kisses and call me 'darling' before I left the room. I think she takes the fondness of the 'intercourse' as an important factor in establishing its success. But she's also an accomplished woman in a rush, and was in and out of the house quickly. I would vacate the bedroom and she, with her herbal tea in hand, would go up to the bedroom and inseminate herself. After two or three failed attempts she consulted some lesbian friends, who are very proficient with the turkey-baster, to ensure her technique was right. By the third month she'd bought a longer turkey-baster.

Not that I ever saw it, but her procedure was to inject herself, stimulate herself sexually for a better result, and then finish it off in meditation with her legs in the air, which I thought was a great accomplishment! But it didn't take too long, because she was out the door and off to a power breakfast by 8.30 am.

After a time, we gave up and went to the IVF clinic to try IUI. I had to have my sperm tested and incubated for six months and then tested again.

Just as they were about to proceed after the six-month waiting period, they suddenly discovered I needed to have other round of genetic tests. Poor Janine was told she had to wait another whole month. The excuse was it had slipped the doctor's mind, because this was a new law or regulation. And then it happened a second time. After waiting another month, they remembered yet another test I had to do.

I thought it was pure incompetence. The two of us went to see the doctor, who was extremely nervous and banging around papers and folders and talking about being a scientist who tries innovative principles that are always successful. I leant forward across his desk and said, 'I would like you to be a bit less the scientist and a bit more the communicator. Now, can we go through some basic questions?'

I said, 'Do I need to throw a brick through your window in the name of gay liberation? Why is this happening?' He said the genetic tests for sperm donors were part of a new government law for couples who are not sexually connected. Then I suddenly understood the reason for the tests. If they were going to introduce my sperm to Janine for the first time, they would be legally liable if that semen was a carrier of disease. It's not so much the nature of my sexuality; it's the lack of sexuality between Janine and me. I was sympathetic to that, but maybe I've got too conservative in my old age and should have thrown a brick through his window anyway!

My emotional investment in fatherhood is dwarfed by the intensity with which Janine desires a baby and what she's going through in the clinical process. When we met she was in the process of choosing a sperm donor and I said, 'Hold going to the donor for a moment and talk to me for a while. I think I could contribute much more than a donor.' I really believe that. I'm contributing a culture and a family. I'm contributing love, stimulation, and an ambition for the child to experience many things. I'm also offering the opportunity to be loved by me and the many people in my life who love me and will love the child as well—namely Tim, but also my family, although we're all getting on a bit. I have projected often that I'll be 70-something on the child's 21st birthday. Sadly, the child will not know my kind and wonderful parents, who've both died in the past ten years. They gave me empathy, filial love, and a feeling of belonging.

I'm also pretty self-loving in a self-doubting kind of way. I also have a quest to experience unconditional love. My mother adored me as the favourite of five children — and, remarkably, I get on very well with my other four siblings — but she, like all of us, was a very conditional giver of love. I feel I have that quality, too, and probably love my own partner conditionally. I want the experience of being consumed by the unconditional love that only your own child can inspire.

I want something to take me away from the real world of ambition. I want something else for the later years in my life. When I think about it, we should be all having children later in life — when we have time, peace of mind, the security of income, and don't have the distraction of careerism, which is now waning for me. Now is a good time for fatherhood, because I've got the peace, space, and wealth to be able to do it without the struggle.

To me, there are many benefits of having a gay father — and particularly this one, in this relationship. I don't want to make a social experiment out of a child or give him or her all that baggage that a gay lib-speaking father might be tempted to do. But I think gay people who have come through the experience of coming out and forging their own identity and priorities are more remarkable people than those who haven't had to go through that experience. I'm not arguing for gay superiority, but people who have had to question things that heterosexual people don't — such as whether to take the risk of holding their partner's hand on an Adelaide street in the 1970s — can come through the other side with compassion, which makes them stronger and wiser people.

I don't quite yet know how we'll negotiate the potential prejudices the child might face because of my sexual orientation, but I wouldn't not do this because of those attitudes; otherwise I wouldn't exist nor would my relationship with Tim. How this child deals with that battle, I'll work out at the time, with Janine.

It seems like a tired old cliché, but I actually have to say I'd prefer to have a boy. My younger brother, whose own daughters I adore, would be furiously jealous if I have a boy. I'd upstage him by having a son to bear our surname.

Whether I'd prefer my child to be heterosexual or homosexual is another question. It's obviously easier in life to be heterosexual. It would therefore be easier for my child to turn out heterosexual, and would bury all sorts of stupid myths about a homosexually reared child. I could just see my own family, who are very supportive of my relationship and me, rolling their eyes and saying, 'I told you so!', if the child turned out gay.

I don't know how to express my disappointment at our failed attempts compared to Janine, who has been so devastated. I realise I will never be as profoundly distressed as her if we're not ultimately successful. I'll be disappointed and mournful of the lost opportunity; but I never expected it, unlike most heterosexual women who probably expect to get pregnant eventually. This feeling has come much later in life to me.

I'll leave it to Janine to decide how long she wants to keep trying. Some of my bolshie friends have given the clichéd male response, 'It's time to get a younger woman.'

But I'm just not that desperate. I won't seek out someone else to have a child with, because I'm not prepared to have a child with any woman, in any circumstances. I like Janine's hard-working, migrant background. I like her flair and attractiveness. I could see it working quite well. I was enjoying the image of us growing closer in a gradual way.

Some people get their identity and purpose from parenthood or to cover up vacuums in their relationships. When you don't have children you forge your own life with all its freedoms, as well as challenges of identity and self-worth. Someone to carry on my genes is an attraction, but it's not a motivation for me.

Glossary
Edited by Dr Lina Safro (MD)

Adhesions
Scar tissue found between abdominal or pelvic organs, which can interfere with the fallopian tubes. Adhesions can be caused by infections, endometriosis, or previous operations.

Amniocentesis
A procedure in which a small amount of amniotic fluid is removed via a needle from the sac around the foetus. The fluid is studied for chromosomal and other abnormalities that may affect the health of the baby.

Assisted reproductive technology (ART)
General term referring to artifical methods used to achieve pregnancy. It includes, among other techniques, taking medications to induce ovulation or in-vitro fertilisation.

Balanced chromosomal translocation
Chromosomes occur in pairs in all cells except sperm and eggs. If part of one chromosome is found connected to a completely different chromosome it's 'translocated'. For that person there

is no net gain or loss of genetic material, so the translocation is 'balanced' and there is no problem. But when that person makes eggs or sperm, some of these will have too much or too little genetic material. The same will be true for any embryo that results. The chromosomal translocation will then be 'unbalanced' and the embryo will usually miscarry.

Blastocyst
An embryo, approximately five days old, consisting of some 100 cells that form an outer layer surrounding a fluid core. It is the last stage of development an embryo reaches before implanting itself into the uterine wall. Transferring blastocyst-stage embryos can increase the odds of implantation.

Blighted ovum
A blighted ovum (also known as an 'embryonic pregnancy') happens when a fertilised egg attaches itself to the uterine wall, but the embryo does not develop. Cells develop to form the pregnancy sac, but not the embryo itself. A blighted ovum usually occurs within the first trimester before a woman knows she is pregnant.

Cervix
The opening between the uterus and the vagina. Cervical mucus plugs the cervical canal to prevent foreign materials from entering the reproductive tract. The cervix remains closed during pregnancy and dilates during labor and delivery to allow the baby to be born.

Cervix, incompetent
A weakened cervix, which opens prematurely during pregnancy and can cause the loss of the foetus. A cervical cerclage is a procedure in which a temporary stitch or two is put around the cervix to prevent its opening until the pregnancy is to term.

Chorionic villus sampling (CVS)
A test in which a needle is passed into the developing placenta (or chorion). A few small fragments of tissue are then tested to show whether or not the developing foetus has certain abnormalities. Also known as placenta biopsy or placentocentesis, it is usually offered to women aged 37 or over. CVS can be performed at an earlier stage of pregnancy than amniocentesis.

Chromosome
Thread-like structures found in the nucleus of all cells (except red blood cells), which contain genetic material (DNA). Chromosomes come in pairs, and a normal human cell contains 46 chromosomes — 22 pairs of autosomes and two sex chromosomes.

Deoxyribonucleic acid (DNA)
A nucleic acid that contains the genetic instructions or the blueprint for the development and function of all living things. The DNA segments that carry genetic information are called *genes*.

Dilatation and curettage (D&C)
An operation in which the cervix (neck of the uterus) is dilated and the endometrium (lining of the womb) is scraped off with a curette. A D&C is often performed on a recently pregnant woman who has miscarried, to remove tissue remaining in the uterus, treat abnormal bleeding, or to obtain a specimen for diagnostic purposes.

Donor egg
In this procedure, eggs are obtained from the ovaries of another woman (donor), fertilised by sperm from the recipient's partner, and the resulting embryos are placed in the recipient's uterus. Older women and women who experience premature menopause are candidates for egg donation.

Ectopic pregnancy
Any pregnancy that implants in a site other than the uterus. The cause is often unknown, but tends to occur when the fallopian tube is damaged in some way.

Egg retrieval (or egg harvest or pick-up)
A procedure used to obtain eggs from ovarian follicles for use in in-vitro fertilisation. The procedure may be performed during a laparoscopy or by using a long needle and ultrasound to locate the follicle in the ovary.

Ejaculation
The physiological process by which semen is propelled from the testicles, through the reproductive tract and out the opening of the penis.

Embryo
The early product of conception; the undifferentiated beginnings of a baby; the conceptus.

Embryo cryopreservation
A method for freezing embryos at a very low temperature to keep them viable, with the intention to thaw and transfer them to the uterus at a later date.

Embryo donation
The transfer of an embryo resulting from egg and sperm that did not originate from the recipient and her partner.

Embryo transfer
Placing an egg fertilised outside the womb into a woman's uterus or fallopian tube.

Embryo transfer cycle
An ART cycle in which one or more embryos are transferred into the uterus or fallopian tube.

Endometriosis
A condition in which endometrial tissue (tissue that lines the inside of the uterus) grows outside the uterus. Endometriosis can cause painful menstrual periods, abnormal menstrual bleeding, and pain during or after intercourse. It may be treated surgically, with medications, or a combination of both.

Fallopian tubes
Ducts through which eggs travel to the uterus once released from the follicle. Sperm normally meet the egg in the fallopian tube, the site at which fertilisation usually occurs.

Follicle
Fluid-filled sac in the ovary that contains a developing egg.

Follicle-stimulating hormone (FSH)
A hormone secreted into the bloodstream that controls the growth of eggs, or oocytes, and the production of the female hormone, oestrogen, by the ovaries. By the time a woman reaches menopause, the remaining eggs in the ovaries do not respond well to FSH, and the level of FSH in the bloodstream increases in an attempt to stimulate the ovaries.

Gamete
A reproductive cell (eggs in a woman and sperm in a man).

Gamete intrafallopian transfer (GIFT)
A treatment for unexplained infertility that is identical to IVF up to the stage of egg retrieval. Instead of incubation of eggs and sperm (gametes) in the laboratory, the gametes are laparoscopically

injected into the fallopian tubes on the day of collection.

Genes
The basic biological units of heredity composed of DNA.

Human chorionic gonadotropin (HCG)
The presence of this hormone in the woman's blood or urine represents a positive pregnancy test. HCG is also given as a medication in an ART cycle to trigger egg maturation and to mimic a luteinising hormone surge.

Hyperemesis
Around three in every thousand pregnant women experience severe nausea and vomiting known as hyperemesis gravidarum (HG). A woman with HG generally has no respite from severe nausea, vomits many times each day, and finds that eating or drinking makes her feel worse, leading to weight loss. If this continues without treatment, she becomes dehydrated and malnourished.

Hysterosalpingogram (HSG)
A method of taking x-rays while injecting dye through the uterine cavity and fallopian tubes. It can reveal blocked or dilated fallopian tubes, fibroids or adhesions within the uterine cavity, or an abnormally shaped uterus. The test is usually performed on days seven to ten of the menstrual cycle.

Hysteroscopy
A hysteroscopy is an examination of the inside of the uterus through a fiberoptic telescope inserted through the vagina and cervical canal. It helps determine the presence of fibroids, polyps, scars, or other abnormalities within the uterus.

Infertility
A condition of the reproductive system that impairs one's ability to conceive children. Conception is a complex process depending on many factors. When just one of these factors is present, infertility can result.

Intracytoplasmic sperm injection (ICSI)
A form of micromanipulation which involves the injection of a single sperm directly into the mature egg (oocyte). The process increases the likelihood of fertilisation when there are abnormalities in the number, quality or function of the sperm.

Intrauterine device (IUD)
A plastic or metal device inserted into the uterus (womb) to prevent conception. Also known as the coil or loop because of the way it was originally shaped.

Intrauterine insemination (IUI)
The insertion of sperm that have been carefully washed and prepared directly into the uterus at the time of ovulation. IUI is usually performed in couples with male-factor infertility such as low sperm concentration or decreased sperm motility.

In-vitro fertilisation (IVF)
A method of assisted reproduction in which the man's sperm and the woman's egg (oocyte) are combined in a laboratory dish, where fertilisation occurs. The resulting embryo is then transferred to the uterus to develop naturally. Usually, one to two embryos are transferred each cycle. IVF can be used to treat women with blocked, damaged, or absent fallopian tubes. It may also be used to overcome infertility caused by endometriosis, male-factor infertility, or unexplained infertility.

Implantation
The embedding of the embryo into tissue so it can establish contact with the mother's blood supply for nourishment. Implantation usually occurs in the lining of the uterus; however, in an ectopic pregnancy it may occur elsewhere in the body.

Impotence
The inability of the man to have an erection and to ejaculate.

Laparoscopy
An operation performed under general anaesthesia whereby a surgical instrument and camera are inserted through a small incision below the navel to view the ovaries, fallopian tubes, and uterus. Other incisions may also be made through which additional instruments can be inserted to diagnose and treat pelvic disease.

Luteinising hormone (LH)
This hormone is released from the woman's pituitary gland, into her bloodstream, triggering ovulation.

Miscarriage
The spontaneous loss of an embryo or foetus from the womb.

Missed miscarriage (or 'chemical pregnancy')
This occurs when the foetus dies but the woman's cervix stays closed, there is no bleeding, and the foetus continues to stay inside the uterus. Some people also refer to this as a 'silent miscarriage'. A missed miscarriage is not usually discovered until several days or weeks after the foetus has died.

Obstetrician
A doctor who specialises in pregnancy and childbirth.

Ovarian cyst
A fluid-filled sac inside the ovary. An ovarian cyst may be found in conjunction with ovulation disorders, tumors of the ovary, and endometriosis.

Ovarian failure
The failure of the ovary to respond to FSH stimulation from the pituitary because of damage to or malformation of the ovary. Ovarian failure is diagnosed by elevated FSH in the blood.

Ovulation
The release of the egg (ovum) from the ovarian follicle.

Petri dish
A shallow, round transparent dish, with a lid, used for growing microbes.

Placenta
The embryonic tissue attached to the uterine wall that exchanges the baby's waste products for the mother's nutrients and oxygen. The baby is connected to the placenta via the umbilical cord.

Placenta praevia
In most pregnancies, the placenta implants high in the top of the uterus. Occasionally, part or all of the placenta will implant in the lower segment of the uterus or cover the cervix. In severe cases women are prone to bleed heavily during labour, preventing the baby from being born vaginally.

Polycystic ovaries
A condition found in women who don't ovulate, characterised by excessive production of androgens (male sex hormones) and the presence of cysts in the ovaries. Though PCO can be without symptoms, some possible symptoms include excessive weight

gain, acne, and excessive hair growth.

Preclinical pregnancy (or biochemical pregnancy)
Evidence of conception based only on biochemical data in serum or urine before ultrasound evidence of a gestational sac.

Pre-eclampsia
Also called toxaemia, this is a problem that occurs in some women during the second half of pregnancy. Symptoms include high blood pressure, swelling, and large amounts of protein in the urine.

Preimplantation genetic diagnosis (PGD)
PGD involves testing an embryo that has been created using ART, such as IVF, prior to transferring it to the uterus and allowing it to develop normally. It can be useful for couples who are at risk of having a child with an abnormal genetic condition (such as cystic fibrosis, sickle cell anaemia, and Huntington's chorea).

Progesterone
The hormone produced by the corpus luteum during the second half of a woman's cycle. It thickens the lining of the uterus to prepare it to accept implantation of a fertilised egg.

Semen
The fluid portion of the ejaculate consisting of secretions from the seminal vesicles, prostate gland, and several other glands in the male reproductive tract. The semen provides nourishment, protection, and a medium in which the sperm can travel to the woman's vagina. Semen may also refer to the entire ejaculate, including the sperm.

Semen analysis
A laboratory test used to assess semen quality, quantity,

concentration, morphology (form), and motility. In addition, it measures semen (fluid) volume and whether or not white blood cells, which indicate an infection, are present.

Sperm
The microscopic cell that carries the male's genetic information to the female's egg, the male reproductive cell, and the male gamete.

Sperm chromatin structure assay (SCSA)
A test to measure the level of DNA fragmentation in the sperm, to enhance the diagnosis of and treatment for male infertility. Sperm with high levels of DNA fragmentation have a lower probability of producing a successful pregnancy.

Surrogate mother
A woman who is artificially inseminated and carries to term a baby, which will usually be raised by its genetic father and his partner.

Trisomy
A condition in which three chromosomes rather than a pair occur. Down's syndrome is caused by trisomy, by the addition of an extra chromosome to the 21st chromosome pair.

Ultrasound
A diagnostic technique which uses high-frequency sound waves to create an image of the internal organs. During pregnancy it is used to monitor embryonic development and examine the tubes and uterus.

Umbilical Cord
Two arteries and one vein encased in a gelatinous tube leading from the baby to the placenta. It is used to exchange nutrients

and oxygen from the mother for waste products from the baby.

Urologist
A physician specialising in the genitourinary tract.

Uterus
The hollow, muscular organ that houses and nourishes the foetus during pregnancy.

Vasectomy
The accidental or elective surgical separation of the vas deferens of both testicles as a means of birth control.

Zygote
A fertilised egg which has not yet divided.

Zygote intrafallopian transfer (ZIFT)
An ART in which eggs are removed from a woman's ovaries, and fertilised with the man's sperm in a lab dish, and the resulting embryos are transferred into the woman's fallopian tubes during a laparoscopy.